Vascular Smooth Muscle: Metabolic, Ionic, and Contractile Mechanisms

Research Topics in Physiology

Charles D. Barnes, *Editor*
Department of Physiology
Texas Tech University School of Medicine
Lubbock, Texas

1. Donald G. Davies and Charles D. Barnes (Editors). Regulation of Ventilation and Gas Exchange, 1978

2. Maysie J. Hughes and Charles D. Barnes (Editors). Neural Control of Circulation, 1980

3. John Orem and Charles D. Barnes (Editors). Physiology in Sleep, 1981

4. M. F. Crass, III and C. D. Barnes (Editors). Vascular Smooth Muscle: Metabolic, Ionic, and Contractile Mechanisms, 1982

Vascular Smooth Muscle: Metabolic, Ionic, and Contractile Mechanisms

Edited by

M. F. CRASS, III

C. D. BARNES

Department of Physiology
Texas Tech University Health Sciences Centers
School of Medicine
Lubbock, Texas

1982

ACADEMIC PRESS
A Subsidiary of Harcourt Brace Jovanovich, Publishers

New York London
Paris San Diego San Francisco São Paulo Sydney Tokyo Toronto

ACADEMIC PRESS, INC.
111 Fifth Avenue, New York, New York 10003

United Kingdom Edition published by
ACADEMIC PRESS, INC. (LONDON) LTD.
24/28 Oval Road, London NW1 7DX

Library of Congress Cataloging in Publication Data
Main entry under title:

Vascular smooth muscle: Metabolic, ionic, and contractile mechanisms.

(Research topics in physiology ;)
Includes bibliographies and index.
1. Vascular smooth muscle. I. Crass, Maurice F. II. Barnes, Charles Dee. III. Series. [DNLM: 1. Muscle, Smooth, Vascular. W3 RE488s v. 4 / WE 500 V331]
QP110.V37V36 612'.13 81-17639
ISBN 0-12-195220-7 AACR2

PRINTED IN THE UNITED STATES OF AMERICA

82 83 84 85 9 8 7 6 5 4 3 2 1

Contents

3. Membrane Electrical Activation of Arterial Smooth Muscle

David R. Harder

4. Current Status of Vascular Smooth Muscle Subcellular Calcium Regulation

Julius C. Allen and Richard D. Bukoski

5. The Contractile Apparatus of Smooth Muscle and Its Regulation by Calcium

D. J. Hartshorne

6. Lipid Transport and Atherogenesis: Role of Apolipoproteins

Richard L. Jackson

List of Contributors

Numbers in parentheses indicate the pages on which the authors' contributions begin.

Julius C. Allen (99), Department of Medicine, Section of Cardiovascular Sciences, Baylor College of Medicine, Houston, Texas 77030

Richard D. Bukoski (99), Departments of Physiology and Medicine, Section of Cardiovascular Sciences, Baylor College of Medicine, Houston, Texas 77030

***David R. Harder** (71), Department of Physiology, East Tennessee State University College of Medicine, Johnson City, Tennessee 37614

D. J. Hartshorne (135), Muscle Biology Group, Departments of Biochemistry and Nutrition and Food Science, University of Arizona, Tucson, Arizona 85721

Per Hellstrand (1), Department of Physiology and Biophysics, University of Lund, S-223 62 Lund, Sweden

Richard L. Jackson (163), Division of Lipoprotein Research, Departments of Pharmacology and Cell Biophysics, Biological Chemistry, and Medicine, University of Cincinnati Medical Center, Cincinnati, Ohio 45267

Allan W. Jones (37), Department of Physiology, University of Missouri, Columbia, Missouri 65212

Richard J. Paul (1), Department of Physiology, College of Medicine, University of Cincinnati, Cincinnati, Ohio 45267

*Present address: Department of Physiology and Biophysics, University of Vermont, College of Medicine, Burlington, Vermont 05401.

Preface

Because of the widespread incidence of cardiovascular disease, perhaps no subject of biomedical research is receiving more intensive investigation than the structure and function of vascular smooth muscle. This extraordinary effort has spawned several recent comprehensive and detailed reviews. In accord with the philosophy of the Research Topics in Physiology series, this fourth volume addresses the subject of vascular smooth muscle function by focusing on six selected areas delineated by chapters authored or coauthored by internationally recognized authorities in their specialized areas. In concise fashion, the authors have strived to present the historical backgrounds and theoretical bases of their research areas, placing their work in perspective and identifying directions for future research. Thus, by design, some degree of comprehensiveness and detail is omitted in favor of giving critical overviews of various key areas in the burgeoning study of vascular smooth muscle.

It seems appropriate that the first chapter be a discussion of the complexities of energy metabolism and how metabolic events can be correlated with a simultaneous quantitative assessment of smooth muscle mechanics and the contractile machinery at the molecular level. Chapters 2 and 3 offer the reader a current view of smooth muscle membrane properties in terms of the distribution, transport, and metabolic control of electrolytes and specific aspects of ion conductance and electrical activity. Chapters 4 and 5 are concerned with how smooth muscle cells regulate their contractile activity through regulation of calcium ion fluxes and the interaction, at the molecular level, of calcium ions with regulatory proteins associated with the contractile apparatus.

Each author, in varying extent, makes reference to the relation of possible anomalies in cellular or subcellular smooth muscle metabolic and/or ionic events to the bases of certain types of vascular disease. The final chapter (Chapter 6) is devoted to the events leading to vascular pathology in the form of atherogenesis. The author weaves his concise, expert description of plasma lipoprotein structure, synthesis, and transport among the current concepts of altered vascular smooth muscle lipid metabolism leading to the genesis of atherosclerotic disease.

It is the hope of the Editors that this volume will, in its conciseness and directed discussion, be a unique guide to researchers and clinicians presently engaged in the study of smooth muscle and related areas.

M. F. Crass, III
C. D. Barnes

Vascular Smooth Muscle: Metabolic, Ionic, and Contractile Mechanisms

1

Vascular Smooth Muscle: Relations between Energy Metabolism and Mechanics

Per Hellstrand and Richard J. Paul

I. INTRODUCTION

Vascular smooth muscle (VSM), like all muscle types, can generate force and shorten when excited. While contraction is a fascinating phenomenon in itself, the special adaptations of this muscle type have captured the interest of muscle physiologists. As the primary effector in the regulation of blood flow, VSM, in maintaining vessel caliber against

VASCULAR SMOOTH MUSCLE: METABOLIC, IONIC, AND CONTRACTILE MECHANISMS

ISBN 0-12-195220-7

blood pressure, is called upon to generate large forces for long periods of time. Under similar conditions, skeletal muscle would rapidly fatigue. Furthermore, the maintenance of this force by VSM is carried out with remarkable efficiency. It can be calculated that a vasculature lined with skeletal muscle would require a metabolic input amounting to about twice the organ's entire basal metabolic rate (BMR) simply to maintain vessel caliber—a task which VSM accomplishes utilizing only about 4% of the BMR (Paul, 1980). This economical maintenance of force is accomplished utilizing actin and myosin components of the contractile apparatus that are similar to those of skeletal muscle. In this chapter we will focus on the metabolic and mechanical properties of VSM in an attempt to explore the basis for these specialized characteristics. In recent years, there have been a number of excellent reviews of VSM; in particular, "The Handbook of Physiology" (Bohr *et al.,* 1980) and "Biochemistry of Smooth Muscle" (Stephens, 1977) offer a wide range of comprehensive information. It is not our intent to duplicate the comprehensive reviews of this field given in the above-mentioned works, but rather to focus on recent developments in mechanics and metabolism in an attempt to synthesize new perspectives. We will therefore, because of this emphasis rather than oversight, radically streamline the "review" aspects of this work. We realize that many significant contributions to the field may not receive full acknowledgment in this process. However, we hope that the work will serve as a guide to the literature for those interested in pursuing this field in depth.

II. RELATIONS BETWEEN METABOLISM AND CONTRACTILITY IN VASCULAR SMOOTH MUSCLE

The phrase "vascular smooth muscle," while often used as if representing a homogeneous class, includes many divergent tissues. Differences among vascular tissues are often more pronounced than differences, for example, between cardiac and skeletal muscle. However, over the last decade a fair amount of data on VSM has accumulated, and general patterns of mechanical and metabolic behavior can be discerned.

One of the most obvious metabolic differences between smooth and skeletal muscle is that the phosphagen pool of smooth muscle is 10–20 times lower than in skeletal muscle. The term "phosphagen" is used to describe the chemical substances serving as the immediate source of free energy driving contraction and other energy-requiring processes. These include adenosine triphosphate (ATP) and other so-called high-energy phosphates which can rapidly transfer a terminal inorganic phosphate group (P_i) to adenosine diphosphate (ADP). For example, phosphocreatine participates in the Lohman reaction:

$$\text{Phosphocreatine} + \text{ADP} \rightleftarrows \text{ATP} + \text{creatine}$$

The total phosphagen content of VSM is on the order of 2–4 μmol/g (Paul, 1980) (all weights given are in grams "blotted" or wet tissue weight), which may be compared to a basal rate of utilization of 1–3 μmol g^{-1} min^{-1}. Thus, even under basal conditions, the preformed phosphagen could suffice for only a few minutes in the absence of ATP synthesis via intermediary metabolism. Under conditions of maximum contractile activity, energy demand may increase two-to threefold and, for contraction durations typical of vascular tissues, the preformed phosphagen can provide only a small percentage of the total ATP requirements. In these terms, intermediary metabolism plays a relatively larger role in the mechanochemistry of VSM than in skeletal muscle in which brief contractions are supported entirely from the phosphagen pools, with resynthesis of the ATP utilized usually not occurring until after the contraction is over. On this basis alone one would anticipate a strong relation between metabolism and contractility in VSM. Until the past decade, however, most studies on vascular metabolism ignored contractile conditions entirely. Most experiments were performed on vessel slices, strips, or rings in which the mechanical conditions were unknown or uncontrolled. While these studies are useful for resolving certain qualitative questions, for example, to demonstrate the existence of particular biochemical pathways, the strong dependence of metabolism on contractility tends to obscure quantitative interpretation of such studies.

The development of polarographic electrode techniques for the measurement of oxygen consumption greatly reduced the complexity of simultaneous measurement of oxygen consumption and force. An example of this type of apparatus is shown in Fig. 1. By the end of the 1970s the relation between steady-state oxygen consumption rates (J_{O_2}) and active isometric force (P_0) had been measured for various VSM preparations, including bovine mesenteric vein (Paul *et al.*, 1973), rat portal vein (Hellstrand, 1977), porcine carotid (Paul *et al.*, 1976) and coronary arteries (Paul *et al.*, 1979), and rat aorta (Seidel *et al.*, 1979; Arner and Hellstrand, 1981). From these studies a linear relation between J_{O_2} and P_0 was consistently observed, in spite of the fact that the absolute values of J_{O_2} varied by about fivefold from porcine carotid artery (0.07 μmol min^{-1} g^{-1}) to rat portal vein (0.4 μmol min^{-1} g^{-1}). An example of this dependence is shown in Fig. 2 in which steady-state J_{O_2} is

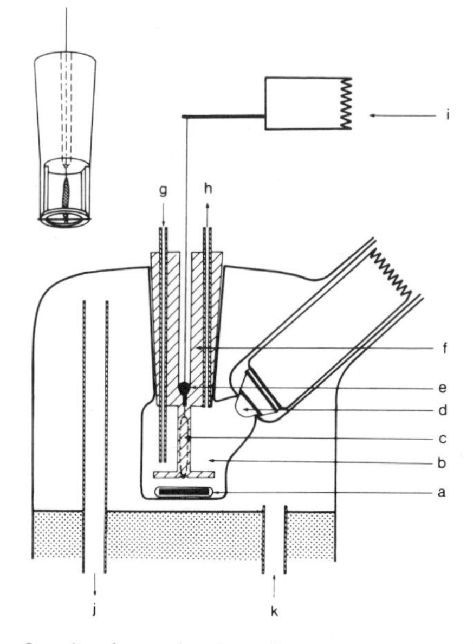

Fig. 1. Apparatus for the determination of oxygen consumption of smooth muscle with simultaneous tension recording. Inset shows muscle holder. a, Magnetic stirrer; b, measuring chamber (volume 1.3 ml); c, muscle preparation (hidden); d, oxygen electrode; e, mercury drop sealing mechanical connection; f, Perspex muscle holder; g, inlet tube for perfusion of chamber; h, outlet tube; i, force transducer; j and k, tubes for circulating water at 37°C. From Hellstrand (1977).

given as a function of isometric force. Based on the time required to attain constant rates of oxygen consumption following changes in contractility, steady states are achieved quickly (<2 min) and can be maintained for many hours provided the tissues are adequately supplied with oxygen and substrate.

Isometric force in smooth muscle can be varied by changing the agonist level in the bathing solution. A linear relation between J_{O_2} and graded isometric force at fixed length has been generally observed and appears to be relatively independent of the agonist studied, including epinephrine, norepinephrine, histamine, and KCl. In an alternative experimental protocol exploiting the force–length characteristic, the agonist concentration may be held constant and the force varied by altering the initial tissue length. Here again, an agonist-independent linear relation has been found; however, this relation, as seen in Fig. 3, is not in general identical to the relation generated at fixed length by changing the agonist concentration. The most common interpretation of these results is dependent on the assumption that a sliding-filament

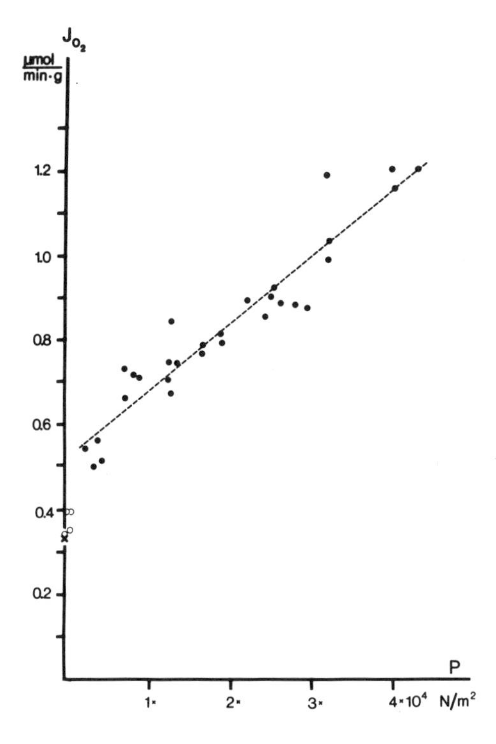

Fig. 2. Oxygen consumption (J_{O_2}) in K^+ contractures in rat portal vein plotted against contracture tension (P). Data from six experiments. Graded contracture force obtained by varying $Ca^{2+}{}_o$. Points obtained with lowest Ca^{2+} concentration (0.05 mM) shown as open circles. Cross symbol shows J_{O_2} in Ca^{2+}-free solution (±SE within size of symbol). Dotted regression line based on points with $Ca^{2+} > 0.2$ mM (solid circles); r = 0.95. $J_{O_2} = 0.51 + 1.6 \times P \times 10^{-5}$, in units given on axes. From Hellstrand (1977).

mechanism, analogous to that operating in skeletal muscle, is valid for VSM as well. As shown in Fig. 3, the linear relation between J_{O_2} and P_0 generated by varying the initial muscle length is a reflection of a near-parallel dependence of J_{O_2} and P_0 on tissue length. Because the dependence of active isometric force on length is most simply interpreted in terms of geometric changes in the number of actomyosin interaction sites, it is natural to view the tension-dependent metabolism, when length is the only parameter altered, as being related to the adenosinetriphosphatase (ATPase) activity of the actomyosin *in vivo*. This view takes further support from the fact that measurements of the actomyosin ATPase *in vitro* are in good agreement with estimates based on the tension-dependent changes in J_{O_2} in the intact tissue (Glück and Paul, 1977; Seidel *et al.*, 1979).

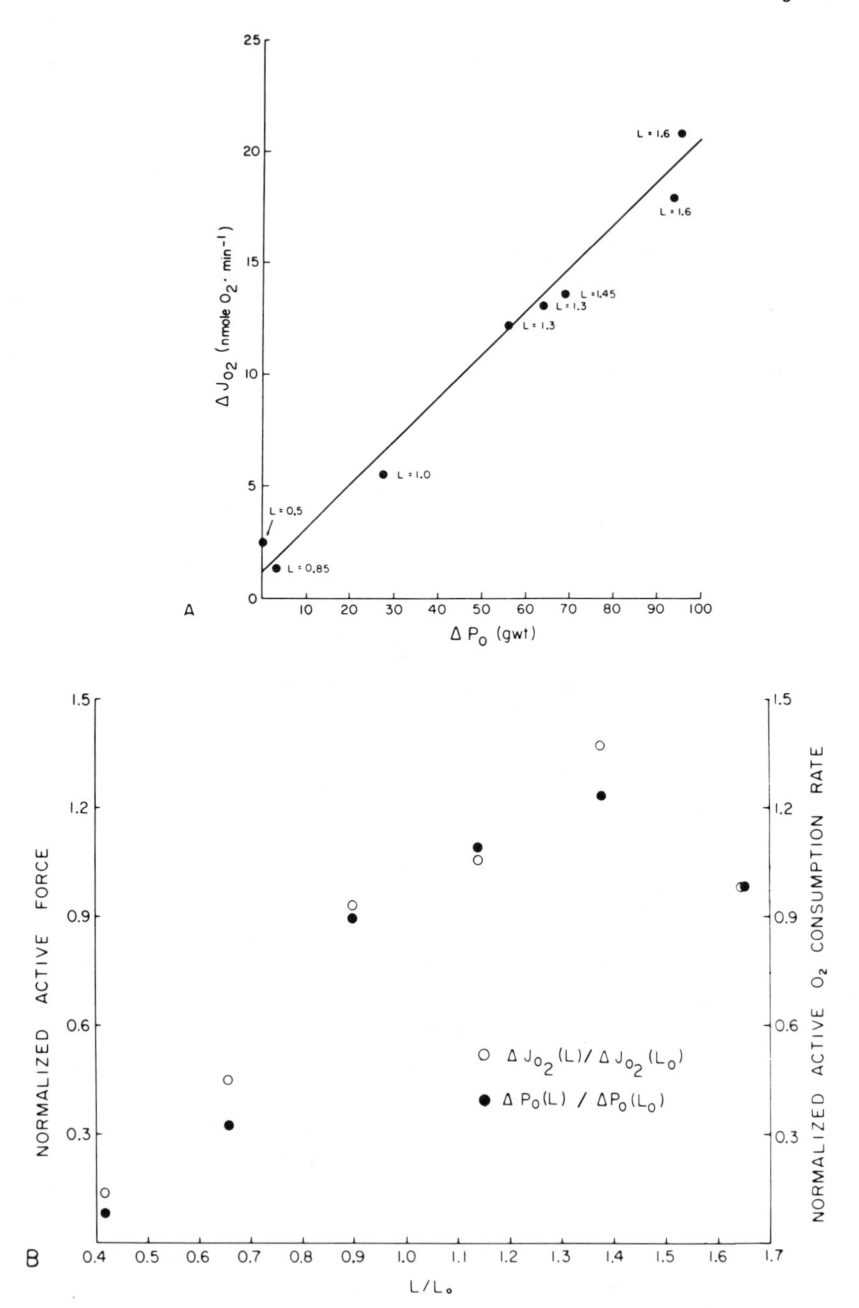

Fig. 3. (A) Suprabasal J_{O_2} plotted against active isometric force from the data shown. Both ΔJ_{O_2} and ΔP_0 for maximal stimulation at varying lengths have been normalized to their

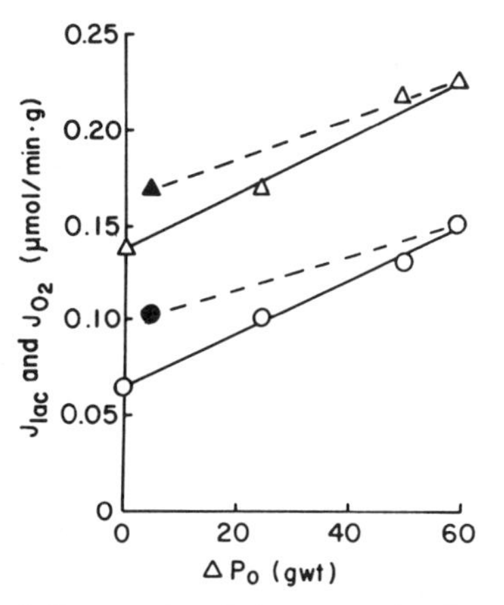

Fig. 4. A plot of metabolic rates in relation to tension in a single segment of bovine mesenteric vein. J_{lac} (triangles) and J_{O_2} (circles) depend linearly on the graded active isometric tension maintained at rest length with different levels of the agonist (epinephrine). At the minimum contracted length (solid symbols), where the developed isometric tension with maximal stimulation is small, both J_{lac} and J_{O_2} are found to be about 20% greater than the respective basal values. From Peterson and Paul (1974).

The relation between J_{O_2} and agonist graded P_0 at a fixed length is generally steeper than that between J_{O_2} and P_0 generated by varying the tissue length (Fig. 4). This difference in metabolic rate may be attributed to non-tension-dependent processes stimulated by the agonist. When interpreted in this manner, about 20% of the suprabasal increase in J_{O_2} observed when maximal isometric force is elicited can be associated with tension-independent processes. The nature of these processes is unclear, but energy-requiring Ca^{2+} translocation and other ion transport processes are natural candidates.

It is of interest to note that, to maintain maximum isometric force, VSM is only required to approximately double its basal J_{O_2}. Skeletal muscle under similar conditions would need to increase its J_{O_2} by approximately 300-fold to provide ATP to match measured rates of ATP

---

respective values at the resting length L_0. Values of length shown with each point are expressed relative to L_0. The standard deviation of the least squares linear regression shown is ±7.5%. The intercept is nonzero and statistically significant. (B) Active isometric force, normalized to the value at resting length, for the maximally stimulated tissue at varying muscle lengths (solid circles) is plotted against muscle length, expressed as multiples of the resting length L_0. A similar dependence on muscle length is seen for the simultaneously recorded values of J_{O_2}, the suprabasal oxygen consumption rate, when normalized to the value of J_{O_2} at the resting length (open circles). All data were taken from a single bovine mesenteric vein segment. From Paul and Peterson (1975).

utilization. This increase well exceeds the maximum oxidative capacity of most mammalian tissues (Hill, 1965). Thus maximal isometric contraction can be maintained in smooth muscle, whereas a contraction of similar magnitude rapidly leads to fatigue in skeletal muscle.

The general pattern outlined above appears to hold for a variety of vascular tissues and is a useful first approximation to the relation between metabolism and contractility in VSM. As with any scientific theory, interest more often lies in the exceptions and the areas not explained by the general observations. It is our goal to explore these facts in the following sections.

III. RELATIONS BETWEEN METABOLISM AND CONTRACTILITY—CURRENT QUESTIONS

A. Substrate for Oxidative Metabolism

Steady-state oxygen consumption is tightly coupled to isometric force. As VSM mitochondria appear similar to mitochondria in general, it is not unreasonable to assume that they are acceptor-limited, i.e., their oxygen consumption is limited by the amount of ADP available. The relation between J_{O_2} and force can thus be ascribed to the increased ADP made available to the mitochondria by the actomyosin ATPase. Because the level of free ADP is very small in tissues, this hypothesis is difficult to test directly, and it is based on changes in ADP calculated from measurements of ATP, phosphocreatine, and creatine, assuming that the Lohman reaction is in equilibrium (Nishiki *et al.*, 1978). Though a reasonable hypothesis based on evidence from other muscle systems, this type of information is not yet available for VSM.

While a generally accepted hypothesis relating mitochondrial ATP output to metabolic demand exists, little is known about the mobilization of substrate for oxidative phosphorylation in VSM. This is in a large part due to a lack of unambiguous knowledge of the nature of the substrate oxidized. Two types of conflicting evidence are available (see Paul, 1980). Measurements of respiratory quotients in vascular tissue cluster near 1.0, suggesting that carbohydrate is a predominant substrate. However, only a small fraction of ^{14}C radioisotope label from glucose is reported to be incorporated into CO_2, implicating some other substrate. Little is known about the effects on the substrate utilization pattern of the level of mechanical activity, a parameter generally not controlled in metabolic studies and potentially responsible for the reported discrepancies. Based on the ability of various substrates to restore isometric force in substrate-depleted preparations, a technique pioneered by Furchgott (1966), the enzyme systems for conversion of carbohydrates, lipids, and proteins into ATP appear to exist in VSM. In the presence of these fuels at physiological levels the preferred substrate (if any) and the effects of contractility remain to be determined. This is not only relevant at the level of basic muscle physiology, for example, in terms of the stoichiometric ratios for calculating J_{ATP} from J_{O_2}, but also at the level of vascular myopathy, in which changes in J_{O_2} may reflect altered substrate utilization patterns.

B. Anomalous Aerobic Glycolysis

While the role of glucose as a substrate for vascular oxidative metabolism remains controversial, there is considerable evidence indicating that glucose transport into VSM is substantial. A long-standing problem for students of VSM, however, is that most of the glucose entering the cell is catabolized only to lactate. The production of lactate under aerobic conditions, known as aerobic glycolysis, has been reported for relatively few cell types such as retinal cells, erythrocytes, and ascites tumor cells. It can be explained in terms of teleological reasoning for some cells, such as retinal cells, in which the presence of oxidative pigments would interfere with light absorbance. Aerobic glycolysis, however, is usually thought of as a defect in cellular metabolism, as some type of failure of the Pasteur effect (Krebs, 1972). In terms of ATP production, about 30% of the total tissue ATP production, at most, is attributable to aerobic glycolysis. However, up to 90% of the glucose utilized by VSM is catabolized by this, relative to the ATP yield of complete oxidation, inefficient pathway. As an example, an increase of only 20% in oxidative metabolism would be needed to replace the ATP provided by aerobic glycolysis under basal conditions. As VSM oxidative metabolism can increase by about 100% under conditions of maximum contractile activity, it is doubtful that aerobic glycolysis can be ascribed to a lack of oxidative capacity in VSM. Other obvious explanations, such as lack of oxygen and nonphysiological experimental conditions, e.g., high bath glucose, have also been shown to be unlikely (Paul, 1980).

Lehninger (1959) suggested that aerobic glycolysis was beneficial in providing a lower pH which prevented $Ca_3(PO_4)_2$ precipitation. However, because significant levels of aerobic glycolysis are not seen in most cell systems, the majority of proposed mechanisms for vascular aerobic glycolysis involve some level of metabolic failure or defect. In fact, the ratio of glycolytic to oxidative metabolism has been suggested as a measure of vascular myopathy (Pantesco *et al.*, 1962; Daly, 1976).

In initial studies relating the rate of lactate production (J_{lac}) under aerobic conditions to isometric force (Peterson and Paul, 1974), J_{lac} appeared to show a relation to isometric force similar to that of J_{O_2}. Figure 4 shows the relation between J_{O_2}, J_{lac}, and P_0 for bovine mesenteric vein. The parallel dependence of J_{lac} and J_{O_2} on P_0, however, was later shown not to hold for all stimulation conditions (Glück and Paul, 1977). Further investigations showed that depolarization achieved by adding KCl to the bathing solution in porcine coronary arteries increased P_0, J_{O_2}, and J_{lac}, whereas stimulation by complete substitution of K^+ for Na^+ increased P_0 and J_{O_2} but decreased J_{lac} below baseline (Paul *et al.*, 1979). This raised the possibility that Na–K transport processes were coupled to aerobic glycolysis, as added KCl is reported to stimulate Na–K transport, whereas removal of all external Na^+ (by substitution of K^+) would be anticipated to inhibit the Na–K pump (Anderson, 1976). This was tested by inhibiting Na–K transport with ouabain or removal of external K^+ ions. In both cases isometric force and J_{O_2} were found to increase, whereas J_{lac} was markedly inhibited. The regulation of lactate production may, however, involve other factors in addition, since K^+-depolarized (20 m*M* Na^+ present) rat aorta shows decreasing J_{lac} with increasing Ca^{2+} concentration and thus contraction, whereas for the rat portal vein the converse is found (see Figs. 9–10). These observations raise many questions. However, they suggest that aerobic glycolysis in VSM may not reflect nonspecific metabolic defects but may be related to Na–K and other ion transport energy requirements. From a biochemical perspective, a control mechanism that allows for independent changes in oxygen consumption and glycolysis is quite interesting. There are precedents for some form of functional compartmentalization of metabolism and transport in other biological systems. A close relation between glycolysis and Na–K transport has been reported for erythrocytes (Solomon, 1978) and also between glycolysis and Ca^{2+} transport by the sarcoplasmic reticulum (Entman *et al.*, 1976). In both of these systems the glycolytic and transport enzymes are postulated to be organized structurally in a large membrane-associated complex, sometimes referred to in the muscle literature as the glycogen particle. Glycogen phosphorylase, a rate-limiting enzyme at the beginning of the glycogenolytic pathway, has been implicated in the coupling of glycolysis and transport by its presence in the glycogen complex. A series of experiments were undertaken to investigate the role of phosphorylase in regulation of the anomalous aerobic glycolysis in VSM.

Table I presents data for the activation of phosphorylase in porcine coronary artery under various conditions. The activity of phosphorylase under stimulation achieved by adding KCl to normal physiological

TABLE I

Glycogen Phosphorylase Activity, Oxygen Consumption, and Aerobic Glycolysis in Porcine Coronary Arteries[a]

Compound added	Percentage change from basal		
	J_{O_2}	J_{lac}	Phosphorylase activity
KCl, 80 m*M*	+68 ± 7	+67 ± 8	+55.8 ± 14.0
Ouabain, 10^{-5} *M*	+21 ± 3.5	−47 ± 5	+40.7 ± 18.6

[a] Total activity in the presence of adenosine monophosphate (AMP) was 0.14 ± 0.02 μmol g^{-1} min^{-1} (n = 20).

saline solution was more than 50% higher than that observed under basal conditions (Paul *et al.*, 1980). This increase paralleled the changes observed in both J_{O_2} and J_{lac}. In the presence of ouabain, an isometric contraction approximately 50% in magnitude of that induced by KCl was elicited. Under these conditions J_{O_2} was also increased, but J_{lac} was inhibited. Phosphorylase activity increased in parallel with J_{O_2}. This suggests that phosphorylase was not directly involved in the observed inhibition of aerobic glycolysis. This evidence shifts one's attention to other control points, for example, the regulation of glucose transport, as potential sites for regulation of aerobic glycolysis. The role of phosphorylase in coordinating metabolism with contractility is unclear. Its activity increases in parallel with isometric force (Namm, 1971; Hellstrand and Paul, 1980), and there is sufficient glycogen in VSM to serve as a substrate for the observed changes in oxidative metabolism. However, the nature of the substrate for oxidative metabolism as outlined above remains ambiguous.

Before attempting to correlate mechanical and metabolic phenomena further, we will review the current status of smooth muscle mechanics in order to provide a sufficient background for further correlations between metabolism and contractility.

IV. MECHANISMS AND ENERGETICS OF CONTRACTION

The focus of this chapter is on the relation between energy metabolism and contractile properties of smooth muscle. Because of such phenomena as spontaneous tone and stress relaxation, the evaluation of contractility in smooth muscle, for which both relaxed and contracted states are often poorly defined, is notoriously difficult. In the last decade, however, considerable progress has been made in characterizing various smooth muscle preparations by the use of concepts originally formulated for skeletal muscle. Accounts of this development are given by Johansson (1975) and Murphy (1976, 1980). We shall, however, consider here some of the results obtained by this approach, since they will be used for our further discussion of chemomechanical transduction. Recent investigations have dealt with the limits of applicability of "classical" mechanics as regards the response of a smooth muscle preparation to very rapid changes in its mechanical constraints, the so-called mechanical transients. This area, again viewed against the background of developments in skeletal muscle physiology, provides new insight into contraction mechanisms at the molecular level and will be considered in conjunction with biochemical evidence on the kinetics of the actomyosin cross-bridge cycle.

A. Classical Analysis of Muscle Mechanics

The view of skeletal muscle mechanics that emerged from the classic studies of A. V. Hill and his school (summarized by Hill, 1970) is often described in terms of a conceptual model in which the muscle is thought to consist of a contractile component (CC) and a series elastic component (SEC), connected as shown in Fig. 5A. The CC is considered freely extensible at rest, but on activation of the muscle it acquires the ability to support a load and, depending on the force acting on it, shortens, remains at constant length, or lengthens. The dynamic properties of the CC are fully described by its characteristic force–velocity relation that may vary however, depending on the length of the CC and the extent to which it is activated (its intensity of active state). Implicit in the concept of the CC is that it has no viscous behavior, i.e., at each instant of time its velocity is determined by the force acting on it only and is not dependent on its history of mechanical perturbations. The SEC of the Hill muscle model is considered to be a purely elastic body that may, however, have nonlinear force–extension properties.

Because a real muscle is not freely extensible but may exhibit a certain degree of resting tension, it is necessary to supplement the two-element model with a parallel elastic component (PEC) that may be arranged in any of the two principal ways shown in Fig. 5B and C. Indeed, to describe the exact mechanical behavior, one may have to use combinations of the two arrangements, even infinite combinations. In most cases,

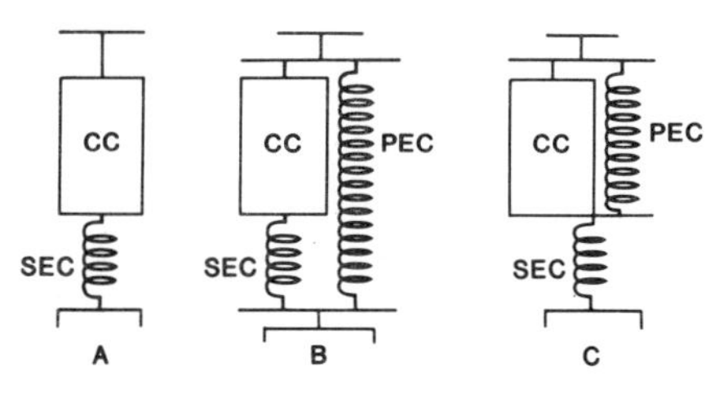

Fig. 5. Different mechanical analogues of muscle. (A) Two-component model; CC, contractile component; SEC, series elastic component. (B and C) Two forms of the three-component model. PEC, Parallel elastic component.

however, mechanical behavior can be adequately described by one of the models in Fig. 5B and C. Particularly in the case of skeletal muscle fibers, the influence of the PEC may be disregarded since resting tension is often small at the optimal length for active tension development (l_0). In muscles where this is not the case, one may nevertheless be able to disregard the PEC by working at muscle lengths below l_0. Within this framework, a complete description of the mechanical properties of a muscle requires information on (1) the applicable arrangement of the CC, SEC, and PEC; (2) the force–extension relation of the PEC; (3) the force–extension relation of the SEC; (4) the force–length relation of the CC; (5) the force–velocity relation of the CC; and (6) the time course of activation of the CC.

Ideally, with this information in hand one is able to infer the force and length of the CC at any instant, as well as the changes in these parameters with time, and the power output (see Jewell and Wilkie, 1958). This information can then be correlated with data on energy turnover obtained by chemical measurements.

B. Characteristics of Elastic and Contractile Components

The force–velocity relation may be observed by the method of afterloaded isotonic contractions (Hill, 1938). However, as discussed by Hellstrand and Johansson (1975) and Murphy (1980), this method implies that the shortening velocity for the various loads will be measured at different points in time and at different CC lengths. This may be a serious drawback in experiments on smooth muscle in which these effects are magnified by its slow tension development and relatively compliant elastic component. One type of experiment that avoids this difficulty and gives information on both SEC and CC properties is an isotonic quick release (Jewell and Wilkie, 1958), whereby the muscle is initially stimulated under isometric conditions and then changed to isotonic conditions in which it may freely adjust its length to a new constant load set by the experimenter. The principal features of the force and length responses in an isotonic quick release are shown in Fig. 6. Coincident with the load step is a change in length, which is abruptly halted as soon as muscle force reaches its new set level. In terms of the Hill two-component muscle model, the initial quick length change is assumed to represent the recoil of the SEC to the step change in force, whereas the subsequent slower shortening is a property of the CC determined by its force–velocity relation. In practice, both of these quantities have turned out to be quite difficult to measure with accuracy, since a sharp force step requires a mechanical system with very low inertia, as well as a force transducer of high-frequency response. These problems have long been recognized by skeletal muscle physiologists, but only recently have experiments on smooth muscle begun to be performed with techniques permitting rapid time resolution. Even so, determination of shortening velocity after a force step in smooth muscle may be ambiguous, as there is a rapid decrease in velocity over the first few tenths of a second (Johansson *et al.,* 1978; Mulvany, 1979). This phenomenon is shown in Fig. 7A and is discussed in more detail later, but it should be pointed out that the values for shortening velocity obtained are quite dependent on how soon after the step they are determined. In practice, one often has to rely on measuring the velocity at a fixed point in time after the release, e.g., 100 msec (Hellstrand and Johansson, 1975).

An alternative method of analyzing dynamic mechanical properties of muscle is to change its length quickly between two set values while measuring the concomitant change in force (cf. Fig. 7B). It is easier to

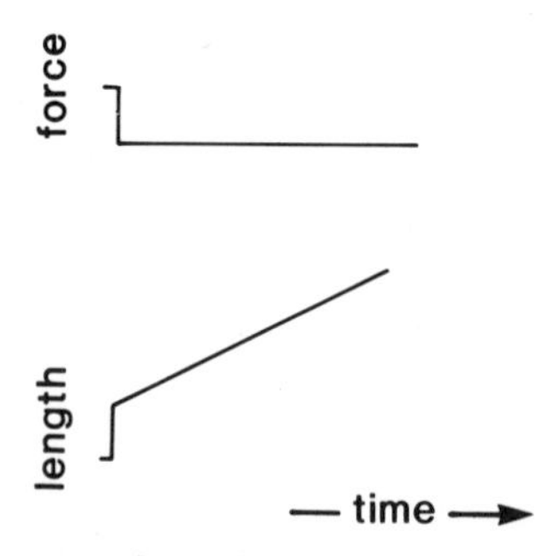

Fig. 6. Force and length response of a muscle to a step change in force. Length record shows shortening as an upward deflection. The figure shows the response expected on the basis of the two-component analogue model. Compare Fig. 7A.

produce a sharp step response in this kind of experiment which, for lack of a better name, has sometimes been called the "isometric quick release" (although this is actually a contradiction in terms). The tension recovery following the step response in the isometric release is not immediately interpretable in terms of the force–velocity relation, although the two kinds of release experiments show corresponding phenomena (Huxley, 1974).

A third way of estimating a muscle's stiffness is to expose it to sinusoidal oscillations of small amplitude and high frequency. This method has been applied to both skeletal and smooth muscle (Julian and Sollins, 1975; Meiss, 1978), and it has been found that force follows length without a significant shift in phase. This indicates that the effects of a viscous component are negligible in this kind of response.

The force–velocity relation of both skeletal and smooth muscle is commonly expressed in terms of the equation

$$V = b(P_0 - P)/(P + a) \quad (1)$$

(Hill, 1938). Here V is the shortening velocity, P the force, P_0 the isometric force, and a and b constants. The force–velocity relation is hyperbolic in form, and it contains information on the power output of the muscle, since this is the product PV. It is found by differentiation that the maximal power is exerted at a force given by

$$P_{\mathrm{m}} = a[\sqrt{(P_0/a) + 1} - 1] \quad (2)$$

P_{m}/P_0 is not strongly dependent on a/P_0; in the region $a/P_0 = 0.15$–0.70 we find $P_{\mathrm{m}}/P_0 = 0.27$–$0.39$. It should be noted that, in addition to the above relation, the power produced by a muscle undergoing substantial shortening is dependent on its force–length relation as well as on parameters relating to its activation. Reported values describing the force–velocity relation for various muscles are given in Table II.

We are now in a position to calculate the power produced by a muscle shortening under the optimal load (P_{m}). This power (Φ_{m}) is given by

$$\Phi_{\mathrm{m}} = ab[\sqrt{(P_0/a) + 1} - 1]^2 \quad (3)$$

Calculated values for various smooth muscles are shown in Table II.

C. Efficiency of Chemomechanical Transduction

Using the principles of thermodynamics one can calculate the minimal chemical reaction rate required to support the power putput of a muscle.

Under the conditions prevailing in skeletal muscle, the hydrolysis of ATP yields a free energy (more precisely, an affinity $A = -dG/d\xi$, where ξ is the extent of reaction) of 54 kJ/mol (Curtin *et al.*, 1974). Though dependent on the exact conditions, this value would not be expected to vary substantially, as the major determinants, for example, ATP, ADP, and Mg^{2+} concentrations, are similar in most cell types. In principle, all this free energy can be converted into work. For example, using the parameters in Table II, rabbit taenia coli could produce a maximum power of 0.18 mW/g at 18°C (extrapolated from 23°C using a Q_{10} of 2.5). At maximum thermodynamic efficiency (the ratio of mechanical work to free energy change due to ATP hydrolysis equals 100%), this would require an ATP hydrolysis rate of 3.4 nmol g^{-1} sec^{-1}. Because the shortening velocity in smooth muscle rapidly decreases with length, the power output in an actual experiment involving shortening would be less than the maximal power given in Table II. For rabbit taenia coli at 18°C, the data of Butler *et al.* (1979) indicate average power outputs of 0.17 mW/g and phosphagen hydrolysis rates of 37 nmol g^{-1} sec^{-1}. This gives an efficiency (that is, work produced per unit of free energy dissipated) of about 9%. This efficiency is considerably less than values calculated from skeletal muscle data. For example, an efficiency of 41% can be calculated in a similar fashion for frog sartorius at 0°C shortening for 1 sec under an optimal load (Curtin *et al.*, 1974). Butler *et al.* also report (1980) that, under conditions designed to maximize work output, the phosphagen breakdown in rabbit taenia coli is about 2½ times that observed for isometric force maintenance. This relative increase in rate is quite similar to that found in skeletal muscle and suggests that the Fenn effect (Fenn, 1923) can also be observed in smooth muscle. Though similar measurements are lacking for other smooth muscles, some interesting comparisons can be made on the assumption that a similar relation between phosphagen breakdown in maximally working and isometric contractions will be found, as in rabbit taenia coli and frog sartorius. On this basis one can calculate that smooth muscle efficiency may vary considerably; for instance, to sustain the maximal power outputs in Table II, the hog carotid artery and the rat portal vein would both maximally require the hydrolysis of about 36 nmol $g^{-1}sec^{-1}$ of phosphagen if working at 100% efficiency at 37°C. If the phosphagen breakdown rate under these conditions were 2½ times the isometric rate, the more slowly contracting carotid artery, with a tension-dependent isometric breakdown rate of 19 nmol g^{-1} sec^{-1} (Glück and Paul, 1977), would be expected to have more than twice the efficiency of the portal vein, with a tension-dependent breakdown rate of 48 nmol $g^{-1}sec^{-1}$ (Hellstrand,

TABLE II

Parameters of the Force–Velocity Relation [Eq. (1)] and Maximal Power [Eq. (3)] for Various Muscles[a]

Muscle	P_0 (mN/mm^2)	V_{max} (l_o/sec)	a/P_0	b (l_o/sec)	Φ_m [$(P_0\ l_o)$/sec]	Φ_m (mW/g)	Reference
Frog sartorius, 0°C	250	1.2	0.26	0.31	1.2×10^{-1}	29.2	Hill (1938)
Hog carotid artery	223	0.12	0.18	0.020	8.8×10^{-3}	2.0	Herlihy and Murphy (1974)
Bovine mesenteric vein	73	0.017	0.15	0.0025	1.2×10^{-3}	0.09	Peterson (1974)
Rat portal vein, K^+	24	0.33	1.27	0.55	7.9×10^{-2}	1.9	Uvelius and Hellstrand (1980)
Rat portal vein, spontaneous activity	19	0.74	0.73	0.54	1.1×10^{-1}	2.2	Hellstrand and Johansson (1975)
Rat mesenteric resistance vessel	204	0.13	0.23	0.031	1.2×10^{-2}	2.4	Mulvany (1979)
Rabbit taenia coli, 23°C	87	0.031	0.33	0.010	3.4×10^{-3}	0.29	Gordon and Siegman (1971)
Dog tracheal	93	0.30	0.23	0.06	2.4×10^{-2}	2.2	Stephens (1973)
Guinea pig taenia coli	150	0.3	0.17	0.05	2.2×10^{-2}	3.3	Mashima and Handa (1969)
Guinea pig taenia coli, 23°C	200	0.10	0.18	0.016	7.0×10^{-3}	1.4	Lowy and Mulvany (1973)
Rabbit urinary bladder	84	0.61	0.16	0.095	4.4×10^{-2}	3.7	Hellstrand and Johansson (1979)
Rabbit rectococcygeus	164	0.25	0.30	0.072	2.5×10^{-2}	4.1	Davey *et al.* (1975)

[a] Values are for 37°C unless otherwise indicated.

1977). Thus there may exist differences among muscle types with regard to their efficiency of chemomechanical transduction. The range of variation in this respect is, however, much smaller than that of the "economy" of tension maintenance, since smooth muscles can maintain isometric force at substantially (50- to 300-fold) less energy cost than skeletal muscle.

D. Mechanical Transients

The Hill analogue model of muscle has continued to be used successfully for many applications, especially those involving cardiac and smooth muscle. Careful investigation of skeletal muscle mechanics, however, has revealed properties inconsistent with this model (e.g., Jewell and Wilkie, 1958). Also, with increased knowledge of the structural organization of muscle, attempts were made to correlate properties of the CC and SEC with the sliding-filament model as formalized by Huxley (1957). Podolsky (1960) showed that the length response of a frog sartorius muscle to a step change in force comprised, following the SEC recoil, a transitional period of changing velocity until the steady isotonic shortening velocity was attained. This behavior was not compatible with the original concept of a highly damped CC. Podolsky and his associates (Podolsky and Nolan, 1972) later extended these observations to isolated muscle fibers and developed an explanation for the transition period of oscillating velocity based on modification of the kinetic parameters in Huxley's (1957) cross-bridge model. Huxley and Simmons (1971) analyzed experiments of the converse kind, i.e., the force response to a step change in length, and observed transient phenomena seemingly similar in origin to those examined by Podolsky and associates (Podolsky and Nolan, 1972; see Huxley, 1974). The model advanced by Huxley and Simmons (1971) assumes that the cross-bridge itself incorporates an elastic element and that it can undergo a series of transitions between discrete states of differing potential energy while remaining attached. These transitions are accompanied by conformational changes resulting in interfilament translation or force development, depending on the external mechanical constraints imposed (i.e., isotonic or isometric contraction, or mixed forms). On the basis of the experiments of Ford *et al.* (1977) the elastic element of the proposed cross-bridge structure is extended by about 6 nm during isometric contraction, i.e., about 0.5% of the half-sarcomere length at l_0. The delayed tension recovery following a step change in force indicates the presence of a moderately damped element of about the same reach within the cross-bridge, so that the total working range of the cross-bridge is 10–15 nm per half-sarcomere.

Rigorous testing of current contraction theories in smooth muscle presents several experimental difficulties, particularly with regard to its mechanical properties. Nevertheless, the investigations that have been performed clearly show that the Hill analogue model is inadequate in explaining the mechanical properties of smooth muscle. This applies to both its SEC and its CC parts.

The large SEC compliance of most smooth muscle preparations probably in part arises from the difficulty in attaching it to the experimental apparatus, since natural connecting structures, such as tendons, are lacking. In several preparations, however, it has been shown that for isometric contraction changes in muscle length and length of individual cells or cell segments are closely proportional (Uvelius, 1976; Driska *et al.*, 1978; Mulvany and Warshaw, 1979), indicating that to a first approximation the mechanical behavior of the entire smooth muscle preparation may reflect the behavior of individual cells. Still, more needs to be known about the transmission of force between contractile filaments and other force-carrying structures in the tissue before interpretation of dynamic mechanical events can be made with certainty.

The Hill model predicts that the force of the SEC is determined by its extension alone (or vice versa), regardless of the state of the CC. However, when length recoils resulting from rapidly imposed force steps were measured in rabbit urinary bladder muscle (Hellstrand and Johansson,, 1979), it was found that, for identical force steps performed in the rising and falling phases of an electrically stimulated short tetanic contraction (<5 sec, 37°C), the length recoils were consistently larger in the falling phase. The difference, out of a total recoil of 4–5%, amounted to a maximum of about 0.5% of the muscle length for releases to a minimal load. In other words, the bladder muscle is "stiffer," where stiffness is defined as $\Delta P/\Delta l$, in the rising part of the contraction. Meiss (1978), on the other hand, using sinusoidal length oscillations, measured greater stiffness of rabbit mesotubarium in the falling phase of contraction. How these results will be reconciled is not clear, but the effects of differences in techniques and preparations cannot be ruled out. An interesting hypothesis in this context is that there may be attached but noncycling cross-bridges in smooth muscle contributing to overall stiffness, as suggested by Siegman *et al.* (1976). This could have important consequences for interpretations of energetic as well as mechanical experiments. In both rat portal vein and rabbit urinary bladder, Uvelius and Hellstrand (1980) found that the maximum shortening velocity was 30% lower in K^+- contractures of 1 to 2-min duration as compared to electrically stimulated phasic (2 to 5-sec) contractions. However, stiffness was not statistically different under either stimulus condition, so if indeed noncycling cross-bridges form an internal resistance in K^+-contractures, this effect can only marginally affect stiffness in these experiments. Peiper (1979) has shown that redevelopment of tension in rat portal vein after inhibition by high-frequency (100-Hz) oscillations occurs more slowly later in a contracture, whether induced by electrical stimulation or pharmacological agents. These results seem to indicate that the duration of activity, rather than the specific stimulus, is correlated with changes in dynamic contractile properties.

There are conditions under which both stiffness and shortening velocity of VSM can show parallel changes. A contraction is elicited in the portal vein and other smooth muscle preparations by exposing the muscle to strongly hyperosmotic solutions (Andersson *et al.,* 1974; McGrath and Shepherd 1976). This type of contraction is characterized by a marked increase in stiffness, as well as a reduction in shortening velocity to about 10% of that in an isotonic medium (Hellstrand and Arner, 1980). The metabolic tension cost, i.e., the ATP turnover per unit isometric force, was found to be similar in contractures elicited by hyperosmolality and high K^+ (Arner and Hellstrand, 1980). A proportionally large population of noncycling cross-bridges in the hyperosmotic medium acting as internal resistance, but not contributing to active force, could explain these phenomena.

The possible existence of attached noncycling cross-bridges has attracted increased interest with the demonstration of a decline in phosphorylation of myosin light chains during the course of a K^+ contracture in swine carotid artery (Driska *et al.,* 1981). This decline from an initial high value early in the contracture correlates temporally with a decline in shortening velocity, whereas isometric force maintenance is not affected (Murphy *et al.,* 1980). This raises the possibility that the state of myosin light-chain phosphorylation may control the kinetic properties (for example, the cycle time) of cross-bridges, in addition to permitting actin–myosin interaction. Determination of the phosphorylation state of myosin light chains under the hyperosmotic conditions referred to above would be of particular interest.

The transition from the elastic recoil to the purely isotonic response in VSM occurs with rapidly decreasing velocity (Fig. 7A) (Johansson *et al.,* 1978; Mulvany, 1979). It has been shown in the study by Johansson *et al.* (1978) that this phenomenon is incompatible with a muscle model incorporating a CC whose velocity is dependent on load only. Rather, the effect arises because of the perturbation introduced by the force step and may thus be regarded as a viscous property of the muscle. This viscoelastic transient differs from the effect seen in frog muscle in that the length change after the force step is monotonic, both for decreases

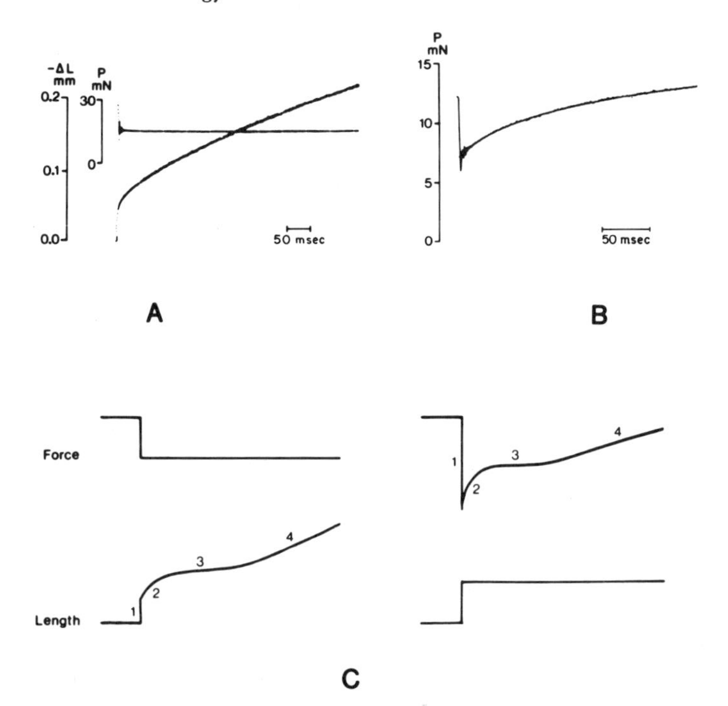

Fig. 7. (A) Oscilloscope recording from a force step experiment on an ac-stimulated preparation of urinary bladder smooth muscle. Upper sweep shows force, and lower sweep length change. The most rapid responses, not clearly visible on the film, have been indicated by dotted lines. The muscle (initial length 4.6 mm, weight 1.7 mg) was stimulated for 3 sec, and the release was triggered at the peak of the isometric contraction, giving a force step from 27 to 16 mN. The change in length is seen to consist of a rapid recoil and a subsequent shortening, the volocity of which decreases rapidly within the first 50 msec and more slowly thereafter. (B) A plot of the force recovery after a shortening step of 1.2% of the muscle length during contraction in a strip of rabbit urinary bladder. Data were sampled at a frequency of 1 kHz and are displayed together with a fitted double exponential (broken line, not clearly distinguished from the sample values). The amplitudes and time constants of the two exponential components are 1.4 mN, 26 msec and 6.2 mN, 196 msec, respectively. (C) Schematic drawing of mechanical transients as described in frog skeletal muscle (Huxley, 1974). (Left) "Velocity (or isotonic) transient." (Right) "Force transient." To conform with usage in (A), shortening is indicated as an upward deflection of the length record. In the model proposed by Huxley and Simmons (1971; Huxley, 1974) phase 1 is attributed to elastic recoil in attached cross-bridges, phase 2 (1–2 msec at 0°C) to a viscoelastic relaxation in attached bridges, phase 3 (5–20 msec) to adjustment of the rates of attachment and detachment of bridges after the step, with the change in detachment rate coming to an end first. Phase 4 represents shortening, or tension redevelopment, respectively, with attachment as the predominant process. From Hellstrand (1979).

and increases in tension, whereas in skeletal muscle oscillatory responses are often seen (Fig. 7C). The force response to a step change in length in smooth muscle is monotonic as well (Fig. 7B) (Hellstrand and Johansson, 1979), which contrasts to the situation in skeletal muscle in which a plateau or even a direction change in the force response is seen after the first recovery (Fig. 7C).

The most prominent difference between smooth and skeletal muscle responses, however, is in the time scale. The length transient in smooth muscle of rabbit urinary bladder was analyzed by Hellstrand and Johansson (1979) as a decaying exponential process with an amplitude of up to 1–1.2% of the muscle length and a time constant of 15–30 msec. Similar results have been obtained from rat portal vein (Uvelius and Hellstrand, 1980), whereas in rat mesenteric resistance vessels the time course of the response appears to be considerably slower although otherwise similar (Mulvany, 1979). For a comparison with skeletal muscle data, it should be noted that in frog muscle at 0°–2°C the response consists of two distinct phases: quick recovery and plateau (cf. Fig. 7C). Whereas the quick recovery occurs in only a few milliseconds, the total response spans a time interval of 10–25 msec (Huxley, 1974). This duration in skeletal muscle is about one-fourth of the total duration of the transient observed in bladder muscle (evaluated as three times the time constant). Thus it appears that the transients observed in smooth muscle may be comparable to the total response in skeletal muscle if the differences in shortening velocity are taken into account (see Section III,E). The fact that two distinct phases are not observed in the smooth muscle transient may be related to the difficulty in obtaining homogeneous responses in multicellular preparations, but could also indicate distinctive features of the kinetics of actomyosin interaction in smooth muscle.

E. Implications for the Molecular Mechanism of Contraction

Based on foregoing metabolic and mechanical data, it is possible to draw some interesting conclusions about the nature of mechanochemical energy transduction by VSM. The tension cost is 10 times less in porcine carotid artery at 37°C (Paul *et al.,* 1976) than the steady value in frog sartorius at 0°C (Paul and Kushmerick, 1974), and nearly 300 times less when compared at the same temperature (Table III). Rat portal vein, a VSM with myogenic phasic activity, has a considerably higher V_{max} than a tonically contracting porcine carotid artery and a tension cost approximately 20 times higher (Hellstrand, 1977), although 15 times smaller than that of the frog sartorius at the same temperature. The range

TABLE III

Comparative Aspects of Muscle

	Rat portal vein, 37°C	Porcine carotid artery, 37°C	Frog 0°C	sartorius 37°C[a]
Tension cost $\left(\frac{\mu\text{mol ATP min}^{-1}\text{ g}^{-1}}{\text{mN/mm}^2}\right)$	0.12	0.006	0.06	1.8
Average duration of cross-bridge cycle $\left(\frac{\text{myosin content}}{\text{tension-dependent } J_{ATP}}\right)$ (sec)	0.20	0.75	0.3–0.7	0.01–0.023
Maximal shortening velocity V_{max} (l_o/sec)	0.74	0.12	1.2	36
Maximal force per myosin content (mN/mm^2 ÷ μmol myosin/g media)	0.27×10^4	1.33×10^4	0.18×10^4	0.18×10^4

[a] Values extrapolated from 0° to 37°C, assuming a Q_{10} of 2.5.

covered by the properties of the contractile system in different VSMs is thus large, but in general smooth muscles as a group are characterized by a high economy of tension maintenance when a comparison is made to striated muscle. The basic components of the contractile machinery, actin- and myosin-containing filaments, seem to be quite similar in smooth and striated muscles, although the structural organization of the filaments and the regulation of their interaction are different. Some indication as to the mechanism underlying the high economy of smooth muscles may be gained from estimation of the cross-bridge cycle duration. In view of current muscle theory, one ATP molecule is assumed to be hydrolyzed per cross-bridge attachment–detachment cycle. Based on one cross-bridge per myosin molecule (although two is also a plausible number) and full participation, this duration can be estimated by dividing the myosin content by the tension-dependent rate of ATP hydrolysis. In the porcine carotid artery at 37°C this estimate is 0.75 sec (Paul *et al.*, 1976). In the rat portal vein, using a tension-dependent ATPase rate of 0.077 μmol g^{-1} sec^{-1}/K$^+$ contractures (Hellstrand, 1977) and a myosin content of 0.015 μmol/g (D. Cohen and R. A. Murphy, personal communication), the estimated cycle duration is about 0.20. These values are roughly in the same range as estimates based on frog sartorius data at 0°C of 0.3–0.7 sec (Curtin *et al.*, 1974; Paul and Kushmerick, 1974); a comparison at the same temperature would, however, show a difference by a factor of about 30. This suggests that much of the economy of smooth muscle can be related to its low actomyosin ATPase (and reciprocally, its longer average cross-bridge cycle duration). However, for porcine carotid artery (but seemingly not for rat portal vein) another factor of about 5- to 10-fold would be needed to match cycle duration to tension cost. Such a factor could be a longer effective sarcomere length compared to that of skeletal muscle (Rüegg, 1971); however, this may be straining the approximate nature of this calculation a bit.

Of further significance is the comparison of cross-bridge cycle times estimated on the basis of metabolic measurements to those inferred from studies on rapid mechanical transients. First, it should be noted that the quick recovery phase of 1–2 msec observed by Huxley and Simmons (1971) after a quick (<1 msec) length step was interpreted by them to represent the time during which the cross-bridges initially attached at the time of the length step remained attached before detaching and beginning another cycle. The whole tension recovery and plateau phase of 5–20 msec was thought to represent the time during which the relative rates of attachment and detachment adjusted to the new situation (force recovery) imposed by the length step. We have already seen that the isotonic transients observed in VSM (Hellstrand and Johansson, 1979; Mulvany, 1979) can be placed on this general time scale if adjusted according to the relative shortening velocities. However, it is obvious that the time periods estimated from mechanical transients are about an order of magnitude smaller than the cross-bridge cycle durations estimated from metabolic studies under the isometric conditions cited above. So if mechanical transients indeed convey information about characteristic steps in the cross-bridge cycle, we have to conclude that the period during which the cross-bridge remains attached, i.e., the time when it can actually exert force, is but a small portion of the total cycle duration.

The lifetime of the attached cross-bridge may also be estimated from the isotonic shortening velocity. The maximal shortening volocity (V_{max}) of a frog semitendinosus muscle at 0°C is about 2 l_0/sec (e.g., Edman *et al.,* 1976). On the assumption that the maximal distance over which a single cross-bridge can remain attached corresponds to a displacement of thick versus thin filaments of 15 nm per half-sarcomere (Ford *et al.,* 1977), the maximum duration of the attached part of the cycle under such conditions is about 7 msec (and proportionately longer at lower shortening speeds). Based on the same assumptions for the portal vein, and with a V_{max} at 37°C of 0.33 l_0/sec (K^+ contractures; Uvelius and Hellstrand, 1980), the maximal duration of the attached part of the cycle at V_{max} is about 40 msec. In a similar manner, the attached part of the cross-bridge cycle in porcine carotid artery at V_{max} is 164 msec and in rat mesenteric resistance vessels, 102 msec, based on the data of Herlihy and Murphy (1974) and Mulvany (1979), respectively. In the latter study, Mulvany (1979) reported a duration of about 150 msec for isotonic transients. Thus, the lifetimes of the attached cross-bridges in the different muscles as estimated from shortening velocities are compatible with those suggested by the mechanical transients.

It should be obvious to the reader that these comparisons between skeletal and smooth muscles are very crude and dependent on assumptions about the smooth muscle contractile system that are not founded on solid information. However, on an order-of-magnitude basis, the comparisons are probably valid, and the main point to emerge is that, although the biochemical cross-bridge cycle time seems to be of about the same magnitude in these diverse tissues (given that we choose to compare amphibian skeletal muscle at 0°C with mammalian smooth muscle at 37°C), the part of the cross-bridge cycle spent in the attached state can be very different and typically is longer in muscles with a higher energetic economy of tension maintenance.

These conclusions gain some support from kinetic measurements on smooth muscle actomyosin systems (Mrwa and Trentham, 1975; Marston and Taylor, 1978; Krisanda and Murphy, 1980), which show that the rate-limiting step may involve a long-lived actin–myosin complex. Needless to say, the relation between the total ATPase cycle time and mechanical attachment of actin and myosin can be accounted for by a number of possibilities as to what fraction of the myosin heads are forming attached cross-bridges at any point in time and whether one or several populations of attached cross-bridges, with possibly different kinetic properties, exist. The possibility of detachment of cross-bridges during rapid shortening without ATP hydrolysis, which has been proposed in skeletal muscle (Eisenberg and Greene, 1980; Irving *et al.*, 1981), should also be kept in mind.

V. COORDINATION OF METABOLISM AND CONTRACTILITY

As pointed out in Section II, the metabolic input of ATP must be closely coupled to contractile energy demands because of the relatively low phosphagen pool in smooth muscle. A strong correlation between steady-state metabolic flux and contractility has been shown for many smooth muscles, attesting to this coupling. However, much needs to be learned about the underlying control mechanisms. Of particular interest in this respect is the time course of the change in the rate of oxygen consumption from the basal to the contractile steady state. One current hypothesis previously alluded to is that this coupling is affected by a change in the ADP/ATP quotient brought about by the increase in actomyosin ATPase, although direct evidence for this is lacking for smooth muscle. One trend that appears to emerge is that J_{O_2} early in a contraction is somewhat higher than the steady-state value (Stephens and Skoog, 1974; see also Fig. 8). This may correspond to an increase in ATP and phosphocreatine breakdown during tension generation relative to that measured in the steady state in taenia coli (Siegman *et al.*, 1980). This increased ATPase during tension development, however, may be more related to processes associated with activation than to increased actomyosin ATPase per se, as the breakdown during tension redevelopment following isometric release in the steady state was found not to be greater than that observed during the maintenance of isometric force.

With a fast-responding oxygen electrode and a muscle chamber of minimum volume, the time course of oxygen consumption and force development was measured for a KC1-induced contracture in rat portal vein (Hellstrand and Paul, unpublished observations). The results for one

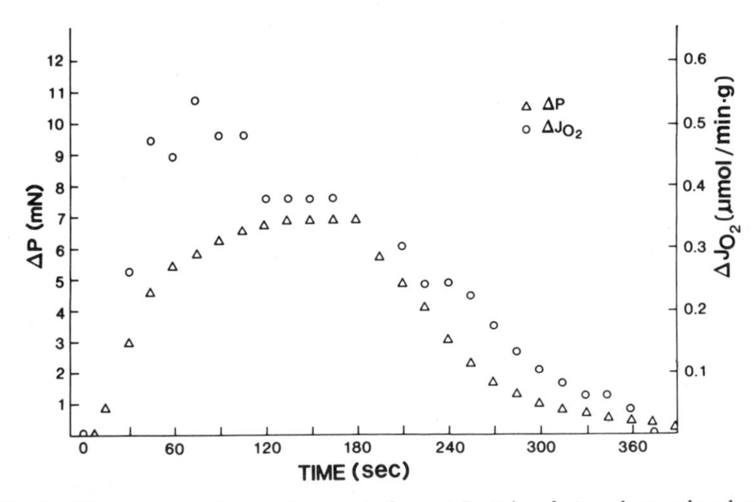

Fig. 8. Time course of active isometric force ΔP_0 (triangles) and suprabasal rate of oxygen consumption ΔJ_{O_2} (circles) in an isometric contracture of rat portal vein at 37°C. The volume of the measuring chamber was 210 μl. The portal vein was first exposed to a high-K^+ depolarizing solution in the absence of Ca^{2+}. At time zero, Ca^{2+} was added to bring the final concentration to 2.5 mM. At 180 sec the depolarizing solution was replaced by normal Krebs solution.

such experiment are shown in Fig. 8. It can be seen that the rate of oxygen consumption changes with a time course that is as rapid as the development of force at this level of resolution. After 1 min of isometric contraction, the level of ATP is not different from that of a resting control, while phosphocreatine is about 0.8 μmol/g less. Whether the change in free ADP associated with the change in phosphocreatine is sufficient to activate mitochondrial oxidative phosphorylation is uncertain and requires further experimentation.

An alternative hypothesis for the activation of metabolism involves intracellular Ca^{2+}. It is attractive in that changes in intracellular Ca^{2+} occur and are known to activate the actomyosin. No known mechanism affecting the coupling of J_{O_2} directly to Ca^{2+}, however, is generally accepted. Nonetheless, there is evidence both for cardiac (Snow *et al.*, 1980) and smooth (Urakawa *et al.*, 1968) muscle that Ca^{2+} may play a direct role in the coordination of oxidative metabolism with contraction. Much experimentation is warranted, and coordination of metabolism and contractility in intact tissues may well be more complicated than can be explained by acceptor (i.e., ADP)-limited mitochondrial respiration.

VI. APPLIED ASPECTS OF SMOOTH MUSCLE MECHANOCHEMISTRY: HYPERTENSION

The contractile energetics of VSM is an interesting and important area in its own right, and the information gained can contribute to comparative studies aimed at revealing general principles of biological energy transduction. Specifically, however, vascular effector organs are involved in the regulation of blood flow and pressure, and disorders in these functions are of great clinical significance.

Physiologists interested in vascular function have directed much of their attention to the study of hypertension, as this disease state can be clearly related to an altered function of the vascular neuroeffector system. Much of the work in the last decade has involved spontaneously hypertensive rats (SHRs), especially the strain developed by Okamoto and Aoki (1963). We will concentrate on this animal model of hypertension, since recent measurements of metabolic turnover in relation to tension development have been reported.

Extensive studies on contractile properties of VSM from SHRs have been performed, and conflicting results obtained as to whether force production in SHRs is greater, smaller, or equal in relation to that in various normotensive controls. However, in the more recent of these studies, where Wistar–Kyoto (WKY) rats have been used as controls, no differences in cellular contractile properties have been found in aorta (Arner and Hellstrand, 1981), portal vein (Mulvany *et al.*, 1980; Arner and Hellstrand, 1981), or mesenteric resistance vessels (Mulvany *et al.*, 1978). It is presently considered that WKY rats, which are genetically more similar to SHRs than either Wistar or Sprague–Dawley rats (e.g., Ooshima, 1973), are the most appropriate controls available. In thoracic aortic media from SHRs and WKY rats, Seidel (1979) found similar total protein, actomyosin, and DNA contents per unit wet weight, even though the total media thickness was greater in SHRs. Since the intraluminal pressure in SHR vessels is higher, there must be either a higher wall tension in these vessels or a smaller diameter (a consequence of the law of Laplace). In fact, it has consistently been found that SHR vessels have a smaller lumen diameter and a greater wall thickness than their normotensive counterparts when a comparison is made at a standarized pressure (Mulvany *et al.*, 1978; Arner and Hellstrand, 1981). This suggests that both of the above possibilities apply. In addition, an increased wall/lumen ratio in hypertensive vessels tends to increase vascular reactivity, i.e., the increase in intraluminal pressure caused by a given rise in smooth muscle contractility (for a review of this concept, see Folkow *et al.*, 1973).

It is of obvious interest, however, to investigate whether the smooth muscle of the vessel wall shows altered properties in hypertension. The above-mentioned data suggest that the mechanical performance of the contractile system of SHR and WKY cells may be similar under isometric conditions, but we would also like to know whether dynamic properties such as stiffness and the force–velocity relation are similar too. Such data are as yet virtually nonexistent, to our knowledge, though Peiper *et al.* (1979) have reported an unchanged V_{max} in portal veins from SHRs compared to those from WKY rats. The portal vein, however, is not exposed to elevated pressure, so investigations on arterial muscle are needed.

A study on isometric force and metabolic rates under these conditions in abdominal aortas and portal veins from SHRs and WKY rats was performed by Arner and Hellstrand (1981). Force was measured in a tangential direction in segments of aorta and in a longitudinal direction in portal veins, the musculature of which is mainly oriented in these respective directions. Oxygen consumption and lactate production were measured in aorta segments under resting conditions and during contraction at l_0 in a high-K^+ medium at two different Ca^{2+} concentrations (0.5 and 2.5 mM); the results were as shown in Fig. 9. Resting J_{O_2} was significantly higher in SHR aortas, and at each tension level this difference persisted to about the same extent, so that the regression line relating J_{O_2} and active stress is displaced upward for SHRs but has the same slope as that for WKY rats. In contrast, no difference between SHR and WKY aortas was seen in J_{lac} (Fig. 9, middle). Interestingly, J_{lac} decreased with increasing Ca^{2+} concentration (and thus tension) in the aortas (Fig. 9, middle) but increased in the portal veins (Fig. 10, middle). The turnover of ATP calculated using standard stoichiometry (cf. Paul, 1980) was greater at all tension levels in SHR aortas (Fig. 9, right). When K^+ contractures in portal veins were studied in a similar way (Fig. 10), no difference in tension-related energy turnover was found between SHRs and WKY rats.

These results can be interpreted to show that arterial muscle from SHRs demonstrates an increased tension-independent rate of metabolism but that the energetics of actomyosin interaction is the same as in WKY rats. However, venous smooth muscle, not exposed to the elevated arterial pressure, does not show a corresponding difference in metabolic rate. Recently, Seidel *et al.* (1980) have also reported an increased BMR in SHR thoracic aortas. Using a protocol different from that of Arner and Hellstrand (1981), they found, however, the tension-dependent metabolism to be higher as well. The nature of these discrepancies is not yet clear.

It is possible that morphological differences can at least in part explain

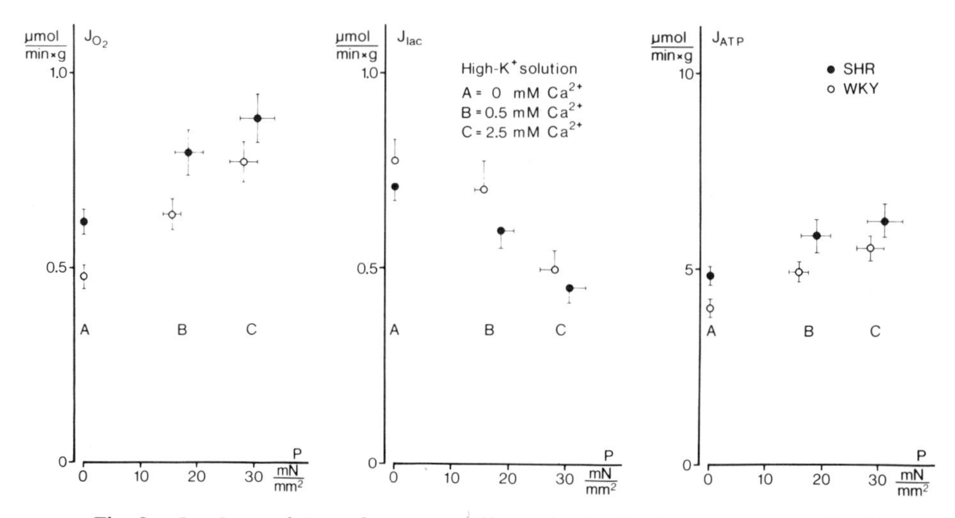

Fig. 9. J_{O_2}, J_{lac}, and J_{ATP} of aortas at different levels of active stress (P). A, B, and C denote 0, 0.5, and 2.5 mM Ca^{2+}, respectively, in high-K^+ solution. Open circles, WKY rats (n = 10); solid circles, SHRs (n = 11). J_{ATP} was evaluated as $6.42 \times J_{O_2} + 1.25 \times J_{lac}$. From Arner and Hellstrand (1981).

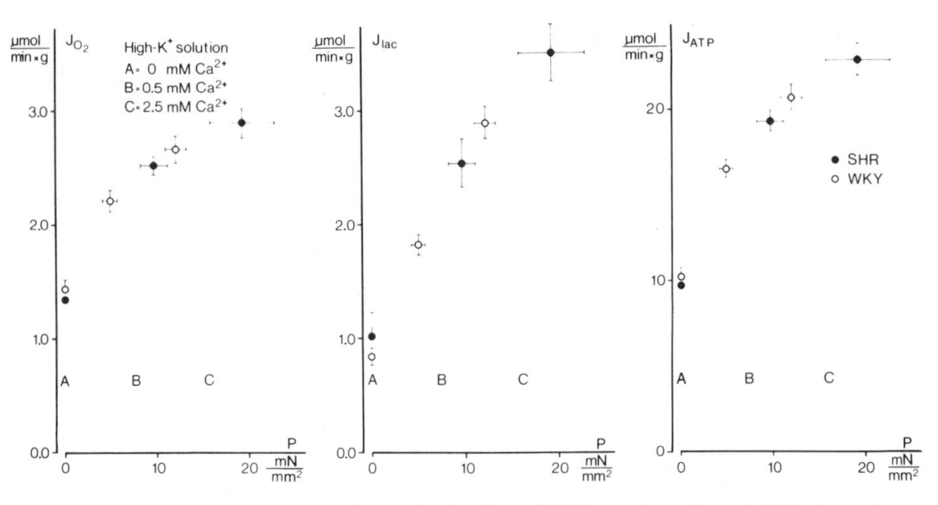

Fig. 10. J_{O_2}, J_{lac}, and J_{ATP} of portal veins from WKY rats (n = 8) and SHRs (n = 9) (open and solid circles, respectively) in high-K^+-induced contractures of different active stress (P). The Ca^{2+} concentration of the medium was 0 mM (A), 0.5 mM (B), or 2.5 mM (C). From Arner and Hellstrand (1981).

the higher metabolic rate of SHR arterial muscle, and this question is currently under study. However, the increased metabolic rate of SHR aorta is interesting in view of reports of increased enzymatic activity related to membrane transport and intermediary metabolism (Ooshima, 1973; Brecher *et al.,* 1978), increased transmembrane ionic flux (Jones, 1974; Friedmann, 1974), and indirect evidence for increased activity of the Na-K pump (Hermsmeyer, 1976; Webb and Bohr, 1979) in VSM from SHR. One would expect such effects to show up as an increased metabolic rate and specifically to affect the tension-independent rate of metabolism. In view of the correlation between aerobic glycolysis and Na-K transport reported for porcine coronary arteries (Paul *et al.,* 1979), one might also anticipate differential effects for oxidative and glycolytic rates.

VII. SUMMARY AND PERSPECTIVES

Within the last decade our knowledge of VSM mechanics, metabolism, and their interrelations has increased dramatically. Oxidative metabolism has been shown to be closely correlated with isometric force, and there are strong indications that the large and puzzling component of aerobic glycolysis is related to Na-K transport processes. The mechanisms underlying this functional compartmentalization and the role played by substrates (which are poorly understood in their own right) supporting vascular intermediary metabolism are unknown and provide challenges for the next decade.

Mechanical studies on VSM have made broad advances based on information gained largely from studies on rapid mechanical transients. These advances, however, have been tempered by a growing awareness of the structural complexity of these tissues compared to skeletal muscle, which makes straightforward interpretations of mechanical transients difficult. There is reasonable evidence attributing part of the stiffness measured in VSM to a property of the cross-bridges themselves. The question of how well this assignment can be quantitated must be answered on a preparation-to-preparation basis. Estimates of cross-bridge properties in intact muscle made from such mechanical measurements as isometric force, stiffness, and contraction velocity will be of particular importance in testing the hypothesis that the state of phosphorylation of smooth muscle myosin light chains controls cross-bridge kinetic properties as well as initiation of actomyosin interaction. It is clear that the technical capabilities are available to make truly quantitative comparisons of VSM mechanics and energetics under normal and pathological conditions. While the results at this point in time are not unambiguous, experiments of this kind can be of substantial value in localizing the cellular defects underlying various vascular disorders.

ACKNOWLEDGMENTS

Original research by the authors was partially supported by the National Institutes of Health (HL23240 and HL22619), the American Heart Association (78-1080), the Swedish Medical Research Council (04X-00028 and 04R-5798), the Medical Faculty of the University of Lund, and AB Hässle, Göteborg, Sweden. During the writing of this chapter, Per Hellstrand was a postdoctoral fellow in the Department of Physiology, University of Cincinnati.

REFERENCES

Anderson, D. K. (1976). Cell potential and the sodium–potassium pump in vascular smooth muscle. *Fed. Proc. Fed. Am. Soc. Exp. Biol.* **35,** 1294.

Andersson, P., Hellstrand, P., Johansson, B., and Ringberg, A. (1974). Contraction in venous smooth muscle induced by hypertonicity, calcium dependence, and mechanical characteristics. *Acta Physiol. Scand.* **90,** 451–461.

Arner, A., and Hellstrand, P. (1980). Contraction of the rat portal vein in hypertonic and isotonic medium: Rates of metabolism. *Acta Physiol. Scand.* **110,** 69–75.

Arner, A., and Hellstrand, P. (1981). Energy turnover and mechanical properties of resting and contracting aortas and portal veins from normotensive and spontaneously hypertensive rats. *Circ. Res.* **48,** 539-548.

Bohr, D. F., Somlyo, A. P., and Sparks, H. V., eds. (1980). "Handbook of Physiology," Sect. 2, Vol. II. Am. Physiol. Soc., Bethesda, Maryland.

Brecher, P., Chan, C. T., Franzblau, C., Faris, B., and Chobanian, A. V. (1978). Effects of hypertension and its reversal on aortic metabolism in the rat. *Circ. Res.* **43,** 561-569.

Butler, T. M., Siegman, M. J., and Mooers, S. U. (1979). Mammalian smooth muscle, economical but inefficient. *Biophys. J.* **25,** 269a.

Butler, T. M., Siegman, M. J., and Mooers, S. U. (1980). Energy utilization in working mammalian smooth muscle. *Fed. Proc., Fed. Am. Soc. Exp. Biol.* **39**(6), 2042 (abstr.).

Curtin, N. A., Gilbert, C., Kretzschmar, K. M., and Wilkie, D. R. (1974). The effects of the performance of work on total energy output and metabolism during muscular contraction. *J. Physiol.* (*London:*) **238,** 455-472.

Daly, M. M. (1976). Effects of age and hypertension on utilization of glucose by rat aorta. *Am. J. Physiol.* **230,** 30-33.

Davey, D. F., Gibbs, C. L., and McKirdy, H. C. (1975). Structural, mechanical, and myothermic properties of rabbit rectococcygeus muscle. *J. Physiol. (London)* **249,** 207-230.

Driska, S. P., Damon, D. N., and Murphy, R. A. (1978). Estimates of cellular mechanics in an arterial smooth muscle. *Biophys. J.* **24,** 525-540.

Driska, S. P., Askoy, M. O., and Murphy, R. A. (1981). Myosin light chain phosphorylation associated with contraction in arterial smooth muscle. *Am. J. Physiol.* **240,** C222-C233.

Edman, K. A. P., Mulieri, L. A., and Scubon-Mulieri, B. (1976). Non-hyperbolic force-velocity relationship in single muscle fibers. *Acta Physiol. Scand.* **98,** 143-156.

Eisenberg, E., and Greene, L. E. (1980). The relation of muscle biochemistry to muscle physiology. *Annu. Rev. Physiol.* **42,** 293-309.

Entman, M. L. Kenichi, K., Goldstein, M., Nelson, T. E., Burnet, E. P., Futch, T. W., and Schwartz, A. (1976). Association of glycogenolysis with cardiac sarcoplasmic reticulum. *J. Biol. Chem.* **251** (10), 3140-3146.

Fenn, W. O. (1923). A quantitative comparison between the energy liberated and the work performed by the isolated sartorius of the frog. *J. Physiol.* (*London*) **58,** 175-203.

Folkow, B., Hallback, M., Lundgren, Y., Sivertsson, R., and Weiss, L. (1973). Importance of adaptive changes in vascular design for establishment of primary hypertension: Studies in man and in spontaneously hypertensive rats. *Circ. Res.* **32/33,** Suppl. I, 2-16.

Ford, L. E., Huxley, A. F., and Simmons, R. M. (1977). Tension responses to sudden length change in stimulated frog muscle fibres near slack length. *J. Physiol. (London)* **269,** 441-515.

Friedman, S. M. (1974). An ion exchange approach to the problem of intracellular sodium in the hypertensive process. *Circ. Res.* **34/35,** Suppl. I, 123-130.

Furchgott, R. F. (1966). Metabolic factors that influence contractility of vascular smooth muscle. *Bull. N.Y. Acad. Med.* **42,** 996-1006.

Glück, E. V., and Paul, R. J. (1977). The aerobic metabolism of porcine carotid artery and its relationship to isometric force: Energy cost of isometric contraction. *Pfluegers Arch.* **370,** 9-18.

Gordon, A. R., and Siegman, M. J. (1971). Mechanical properties of smooth muscle. I. Length-tension and force-velocity relations. *Am. J. Physiol.* **221,** 1243-1249.

Hellstrand, P. (1977). Oxygen consumption and lactate production of the rat portal vein in relation to its contractile activity. *Acta Physiol. Scand.* **100,** 91-106.

Hellstrand, P. (1979). Mechanical and metabolic properties related to contraction in smooth muscle. *Acta Physiol. Scand., Suppl.* **464,** 1-54.

Hellstrand, P., and Arner, A. (1980). Contraction of the rat portal vein in hypertonic and isotonic medium: Mechanical properties and effects of Mg^{2+}. *Acta Physiol. Scand.* **110,** 59-67.

Hellstrand, P., and Johansson, B. (1975). The force-velocity relation in phasic contractions of venous smooth muscle. *Acta Physiol. Scand.* **93,** 157-166.

Hellstrand, P., and Johansson, B. (1979). Analysis of the length response to a force step in smooth muscle from rabbit urinary bladder. *Acta Physiol. Scand.* **106,** 231-238.

Hellstrand, P., and Paul, R. J. (1980). Coordination of metabolism and contraction in rat portal vein: Role of phosphorylase. *Physiologist* **23**(4), 95 (abstr.).

Herlihy, J. T., and Murphy, R. A. (1974). Force-velocity and series elastic characteristics of smooth muscle of the hog carotid artery. *Circ. Res.* **34,** 461-466.

Hersmeyer, K. (1976). Electrogenesis of increased norepinephrine sensitivity of arterial vascular muscle in hypertension. *Circ. Res.* **38,** 362-367.

Hill, A. V. (1938). The heat of shortening and the dynamic constants of muscle. *Proc. R. Soc. London* **126,** 136-195.

Hill, A. V. (1965). "Trails and Trials in Physiology," pp. 1-374. Arnold, London.

Hill, A. V. (1970). "First and Last Experiments in Muscle Mechanics." Cambridge Univ. Press, London and New York.

Huxley, A. F. (1957). Muscle structure and theories of contraction. *Prog. Biophys. Biophys. Chem.* **7,** 255-318.

Huxley, A. F. (1974). Review lecture: Muscular contraction. *J. Physiol. (London)* **243,** 1-43.

Huxley, A. F., and Simmons, R. M. (1971). Proposed mechanism for force generation in striated muscle. *Nature, (London)* **233,** 533-538.

Irving, M., Homsher, E. and Wallner, A. (1981). The effect of rapid shortening on ATP utilization and energy liberation of frog skeletal muscle. *Biophys. J.* **33,** 225a(abstr.).

Jewell, B. R., and Wilkie, D. R. (1958). An analysis of the mechanical components in frog's striated muscle. *J. Physiol. (London)* **143,** 575-540.

Johansson, B. (1975). Mechanics of vascular smooth muscle contraction. *Experientia* **31,** 1377-1386.

Johansson, B., Hellstrand, P., and Uvelius, B. (1978). Responses of smooth muscle to quick load change studied at high time resolution. *Blood Vessels* **15,** 65-82.

Jones, A. W. (1974). Altered ion transport in large and small arteries from spontaneously hypertensive rats and the influence of calcium. *Circ. Res.* **34/35,** Suppl. I, 117-122.

Julian, F. J., and Sollins, M. R. (1975). Variation of muscle stiffness with force at increasing speeds of shortening. *J. Gen. Physiol.* **66,** 287-302.

Krebs, H. A. (1972). The Pasteur effect and the relations between respiration and fermentation. *Essays Biochem.* **8,** 1-34.

Krisanda, J. M., and Murphy, R. A. (1980). Tight binding of arterial myosin to skeletal F-actin. *J. Biol. Chem.* **255,** 10771-10776.

Lehninger, A. L. (1959). The metabolism of the arterial wall. *In* "The Arterial Wall" (A. I. Lansing, ed.), pp. 220-246. Baillière, London.

Lowy, J., and Mulvany, M. J. (1973). Mechanical properties of guinea pig taenia coli muscles. *Acta Physiol. Scand.* **88,** 123-136.

McGrath, M. A., and Shepherd, J. T. (1976). Hyperosmolarity: Effects on nerves and smooth muscle of cutaneous veins. *Am. J. Physiol.* **231,** 141-147.

Marston, S. B., and Taylor, E. W. (1978). Mechanism of myosin and actomyosin ATPase in chicken gizzard smooth muscle. *FEBS Lett.* **86,** 167-170.

Mashima, H., and Handa, M. (1969). The force-velocity relation and the dynamic constants of the guinea pig taenia coli. *J. Physiol. Soc. Jpn.* **31,** 565-566.

Meiss, R. A. (1978). Dynamic stiffness of rabbit mesotubarium smooth muscle: Effect of isometric length. *Am. J. Physiol.* **234,** 614–626.

Mrwa, U., and Trentham, D. (1975). Transient kinetic studies of the Mg^{2+}-dependent arterial myosin and actomyosin adenosinetriphosphatases isolated from porcine carotids. *Hoppe-Seyler's Z. Physiol. Chem.* **356,** 255 (abstr.).

Mulvany, M. J. (1979). The undamped and damped series elastic components of a vascular smooth muscle. *Biophys. J.* **26,** 401–414.

Mulvany, M. J., and Warshaw, D. M. (1979). The active tension-length curve of vascular smooth muscle related to its cellular components. *J. Gen Physiol.* **74,** 84–104.

Mulvany, M. J., Hansen, P. K., and Aalkjaer, C. (1978). Direct evidence that the greater contractility of resistance vessels in spontaneously hypertensive rats is associated with a narrowed lumen, a thickened media, and an increased number of smooth muscle cell layers. *Circ. Res.* **43,** 854–864.

Mulvany, M. J. Ljung, B., Stoltze, M., and Kjellstedt, A. (1980). Contractile and morphological properties of the portal vein in spontaneously hypertensive and Wistar-Kyoto rats. *Blood Vessels* **17,** 202–215.

Murphy, R. A. (1976). Contractile system function in mammalian smooth muscle. *Blood Vessels* **13,** 1–23.

Murphy, R. A. (1980). Mechanics of vascular smooth muscle. *In* "Handbook of Physiology" (D. F. Bohr, A. P. Somlyo, and H. V. Sparks, eds.), Sect. 2; Vol. II, p. 325. Am. Physiol. Soc., Bethesda, Maryland.

Murphy, R. A., Aksoy, M. O., and Dillon, P. F. (1980). Regulation in vascular smooth muscle: Ca^{2+} dependent myosin light chain phosphorylation mediates cross-bridge cycling rates. *Fed. Proc., Fed. Am. Soc. Exp. Biol.* **39**(6), 1817 (abstr.).

Namm, D. H. (1971). The activation of glycogen phosphorylase in arterial smooth muscle. *J. Pharmacol. Exp. Ther.* **178,** 299–310.

Nishiki, K., Erecinska, M., and Wilson, D. F. (1978). Energy relationships between cytosolic metabolism and mitochondrial respiration in rat heart. *Am. J. Physiol.* **234**(3), C73–C81.

Okamoto, K., and Aoki, K. (1963). Development of a strain of spontaneously hypertensive rats. *Jpn. Circ. J.* **27,** 282–293.

Ooshima, A. (1973). Enzymological studies on arteries in spontaneously hypertensive rats. *Jpn. Circ. J.* **37,** 497–508.

Pantesco, V., Kempf, E., Mandel, P., and Fontaine, R. (1962). Etudes métaboliques comparées des pariois arterielles et veineuse chez les bovides: Leurs variations au cours du vieillissement. *Pathol. Biol.* **10,** 1301–1306.

Paul, R. J. (1980). The chemical energetics of vascular smooth muscle: Intermediary metabolism and its relation to contractility. *In* "Handbook of Physiology" (D. F. Bohr, A. P. Somlyo, and H. V. Sparks, eds.), Sect. 2, Vol. II, pp. 174–239. Am. Physiol. Soc., Bethesda, Maryland.

Paul, R. J., and Kushmerick, M. J. (1974). Apparent P/O ratio and chemical energy balance in frog sartorius muscle *in vitro. Biochim. Biophys. Acta* **347,** 483–490.

Paul, R. J., and Peterson, J. W. (1975). Relation between length, isometric force, and O_2 consumption rate in vascular smooth muscle. *Am. J. Physiol.,* 915–922.

Paul, R. J., Peterson, J. W., and Caplan, S. R. (1973). Oxygen consumption rate in vascular smooth muscle: Relation to isometric tension. *Biochim. Biophys. Acta* **305,** 474–480.

Paul, R. J., Glück, E., and Rüegg, J. C. (1976). Cross bridge ATP utilization in arterial smooth muscle. *Pfluegers Arch.* **361,** 297–299.

Paul, R. J., Bauer, M., and Pease, W. (1979). Vascular smooth muscle: Aerobic glycolysis linked to Na-K transport processes. *Science* **206,** 1414–1416.

Paul, R. J., Kuettner, C., and DiSalvo, J. (1980). Compartmentalization of oxidative and glycolytic metabolism in vascular smooth muscle (VSM): Role of phosphorylase. *Fed. Proc., Fed. Am. Soc. Exp. Biol.* **39**(3), 581 (abstr.).

Peiper, U., (1979). The rate of post-vibration tension recovery in relation to type and period of activation in the isolated rat portal vein. *Pflügers Arch.* **382,** Suppl. R25.

Peiper, U., Klemt, P., and Popov, R. (1979). The contractility of venous vascular smooth muscle in spontaneously hypertensive or renal hypertensive rats. *Basic Res. Cardiol.* **74,** 21-34.

Peterson, J. W. (1974). Rates of metabolism and mechanical activity in vascular smooth muscle. Ph.D. Thesis, Harvard University, Cambridge, Massachusetts.

Peterson, J. W., and Paul, R. J. (1974). Aerobic glycolysis in vascular smooth muscle: Relation to isometric tension. *Biochim. Biophys. Acta* **357,** 167-176.

Podolsky, R. J. (1960). Kinetics of muscular contraction: The approach to the steady state. *Nature (London)* **188,** 666-668.

Podolsky, R. J., and Nolan, A. C. (1972). Muscle contraction transients, cross-bridge kinetics, and the Fenn effect. *Cold Spring Harbor Symp. Quant. Biol.* **37,** 661-668.

Rüegg, J. C. (1971). Smooth muscle tone. *Physiol. Rev.* **51,** 201-248.

Seidel, C. L. (1979). Aortic actomyosin content of maturing normal and spontaneously hypertensive rats. *Am. J. Physiol.* **237,** H34-H39.

Seidel, C. L., Bauer, M., and Paul, R. J. (1979). Metabolism of the relaxed and contracted rat aorta. *Fed. Proc., Fed. Am. Soc. Exp. Biol.* **38**(3), 1245 (abstr.).

Seidel, C. L., Strong, R., and Calin, S. (1980). Uncoupling of metabolism and force development in hypertensive aorta. *Physiologist* **23**(4), 102 (abstr.).

Siegman, M. J., Butler, T. M., Mooers, S. U., and Davies, R. E. (1976). Cross-bridge attachments, resistance to stretch, and viscoelasticity in mammalian vascular smooth muscle. *Science* **191,** 383-385.

Siegman, M. J., Butler, T. M., Mooers, S. U., and Davies, R. E. (1980). Chemical energetics of force development, force maintenance, and relaxation in mammalian smooth muscle. *J. Gen. Physiol.* **76,** 609-629.

Snow, R. T. Rubanyi, G., Dora, T., Doua, E., and Kovach, A. G. B. (1980). Effect of perfusate Ca^{2+} on the relation between metabolism and mechanical performance in the rat heart. *Can. J. Physiol. Pharmacol.* **58**(5), 570-573.

Solomon, A. K. (1978). Reflections on the membrane-mediated linkage between cation transport and glycolysis in human red blood cells. *Memb. Trans. Processes* **1,** 31-59.

Stephens, N. L. (1973). Effect of hypoxia on contractile and series elastic components of smooth muscle. *Am. J. Physiol.* **224,** 318-321.

Stephens, N. L., ed. (1977). "Smooth Muscle Biochemistry." University Park Press, Baltimore, Maryland.

Stephens, N. L., and Skoog, C. M. (1974). Tracheal smooth muscle and rate of oxygen uptake. *Am. J. Physiol.* **226,** 1462-1467.

Urakawa, N., Ikeda, M., Saito, Y., and Sakai, Y. (1968). Effects of calcium depletion on oxygen consumption in guinea pig taenia coli. *Jpn. J. Pharmacol.* **18,** 500-503.

Uvelius, B. (1976). Isometric and isotonic length-tension relations and variations in cell length in longitudinal smooth muscle from rabbit urinary bladder. *Acta Physiol. Scand.* **97,** 1-12.

Uvelius, B., and Hellstrand, P. (1980). Effects of phasic and tonic activation on contraction dynamics in smooth muscle. *Acta Physiol. Scand.* **109,** 399-406.

Webb, R. C., and Bohr, D. F. (1979). Potassium relaxation of vascular smooth muscle from spontaneously hypertensive rats. *Blood Vessels* **16,** 71-79.

2

Electrolyte Metabolism of the Arterial Wall

Allan W. Jones

I. INTRODUCTION

The electrolyte metabolism of the arterial wall plays an important role in the maintenance of normal excitatory processes. This has been recognized to be of fundamental importance in understanding altered vascular function during hypertension. In fact, many studies on vascular electrolyte metabolism were conducted in order to understand better the role of altered salt metabolism in the pathogenesis of hypertension. These have been discussed in several reviews (Jones, 1981a, 1982; Somlyo and Somlyo, 1970; Tobian, 1960), and therefore the subject of hypertensive changes is not specifically developed in this chapter. Rather, emphasis is placed on the normal electrolyte distribution within the arte-

ISBN 0-12-195220-7

rial wall and the approaches taken in determining the composition of vascular smooth muscle cells. Specific transport mechanisms operative across the smooth muscle membrane are discussed, as is the metabolic control of these processes. It is well recognized that heterogeneities exist in vascular smooth muscle (Bohr and Uchida, 1967; Somlyo and Somlyo, 1968), and these have been reviewed with respect to electrolyte metabolism (Jones, 1980, 1981a). Of the two broad categories of vascular smooth muscle (phasic and tonic), this article focuses primarily on the tonic type, since this type best represents the smooth muscle of the distributing arteries. We have little knowledge of the electrolyte metabolism of microarteries, and a review must await future investigation.

II. ELECTROLYTE DISTRIBUTION

The large extracellular space (ECS) and proportion of the vessel wall composed of connective tissue have made estimates of smooth muscle composition difficult. This is compounded in small arteries by the contribution of other cell types in the adventitia and intima (Garay *et al.*, 1979). Several approaches have been applied to this problem. Although each has shortcomings, where general agreement is seen, one can have some confidence in the results.

A. Chemical Dissection

This approach is based on the rationale that, if one can account for all the noncellular contents, the cellular components can be computed as the difference. That is, total minus extracellular equals cellular. The ECS is heterogeneous, and electrolyte is both dissolved in water and adsorbed to connective tissue, e.g., collagen and glycosaminoglycans. The dissolved component is routinely estimated by the volume of distribution of extracellular markers. Goodford (1968) has shown that the distribution of several markers in the taenia coli of the guinea pig is inversely related to the radius of the molecule. Villamil *et al.* (1968a) has confirmed this observation in the canine carotid artery. It is generally accepted that markers such as inulin and labeled albumin underestimate the extracellular water and that the molecular size of sucrose offers the most reasonable approximation (Jones, 1980). For studies that involve the measurement of ions as well as spaces, a gamma-emitting marker is perferred to ^{14}C- or ^{3}H-tagged saccharides which require sample destruction for liquid scintillation counting. A chelate of ^{60}Co and ethylenediaminetetraacetate (EDTA) produces a marker, ^{60}Co-EDTA, with a volume of distribution similar to that of sucrose or sorbitol (Brading and Jones, 1969; Jones and Swain, 1972). The ^{51}Cr form of EDTA has also been used on vascular preparations (Garay *et al.*, 1979).

The total water, ECS, and estimated cell water are given in Table I for selected arteries incubated under *in vitro* conditions. Analyses of the taenia coli of the guinea pig are included for comparative purposes. The ECS of arteries is a large fraction of the vessel wall (45–60%), while the cellular water is about half this value. The small arteries tend to have more cellular material than the large distributing arteries on the high-pressure side. Visceral smooth muscle of the taenia coli has the largest cell water content of the tissues presented. This reflects major differences in connective tissue content.

Collagen, elastin, and glycosaminoglycans are the principal connective tissue components of arteries. Collagen and elastin contribute most of the dry solids, while glycosaminoglycans are important in the adsorption of cations (Dunstone, 1962). Such adsorption represents a second extracellular phase for electrolytes, which can be estimated from the connective tissue contents. The analyses of collagen, elastin, and glycosaminoglycans appear in Table II. Collagen plus elastin content varies over a twofold range. Arteries with less cell water (Table I) have a higher collagen-plus-elastin value (Table II). This trend extends to the taenia coli which has the highest cell water and lowest collagen-plus-elastin values shown. Glycosaminoglycans constitute about 1–2% of the dry solids. The cellular solids in arteries (total minus connective tissue) vary from 20 to 60% of the total dry solids while the cellular solids represent 90% of the taenia coli. The ratio of cell water to cellular solids is about 3:1 in the taenia coli. A similar or slightly lower ratio is derived for arteries.

The ratio of extracellular water to connective tissue [ECS/(C + E) in Table II] shows a greater range than for the cellular phase. A minimum of 2–3 was observed in the arteries on the high-pressure side, 7 in the pulmonary artery, and greater than 20 in the taenia coli. A ratio of 2.8 was seen in tendons incubated under similar conditions (Brading and Jones, 1969). The density of the extracellular compartment of arteries exposed to high blood pressure and mechanical stress was similar to that for tendon. On the other hand, a more open, less fibrous matrix was present in tissues under the influence of small distending stresses. A similar relation was observed in arteries from puppies taken at different stages of growth and development (Cox *et al.*, 1974, 1976). These differences have functional significance for the adaptive changes in response to altered mechanical stress on the arterial wall.

TABLE I

Water Distribution in Arteries and Taenia Coli[a]

Animal and tissue	Method	No. of experiments	Water (% wet wt) Total	Water (% wet wt) ECS	Water (% wet wt) Cell	Water (kg/kg DS) Total	Water (kg/kg DS) ECS	Water (kg/kg DS) Cell	Reference
Dog	^{60}Co-EDTA	6	74.0	56.1	17.9	2.85	2.16	0.69	Cox *et al.*, 1976
Carotid			± 2.6	± 3.6	± 1.3	± 0.10	± 0.14	± 0.05	
Lingual	^{60}Co-EDTA	10	75.7	46.5	29.2	3.11	1.91	1.20	Jones and Swain, 1972
			± 0.2	± 1.4	± 1.4	± 0.01	± 0.06	± 0.06	
Rat									
Thoracic arota	^{60}Co-EDTA	8	72.6	51.8	20.8	2.64	1.89	0.76	Jones and Hart, 1975
			± 1.6	± 1.6	± 0.5	± 0.06	± 0.06	± 0.02	
Tail artery	[^{14}C]Sorbitol	6	77.2	47.9	29.3	3.38	2.10	1.28	Friedman *et al.*, 1974
			± 1.6	± 0.9	—	± 0.07	± 0.04	—	
Rabbit									
Main pulmonary artery	^{60}Co-EDTA	8	80.2	51.2	29.0	4.06	2.59	1.47	Jones *et al.*, 1973
Guinea pig			± 2.2	± 0.8	± 1.8	± 0.11	± 0.04	± 0.09	
Taenia coli	^{60}Co-EDTA	9	82.7	35.3	47.4	4.78	2.04	2.74	Jones *et al.*, 1973
			± 1.0	± 1.0	± 1.0	± 0.06	± 0.06	± 0.06	

[a] Values are means ± SEM.

TABLE II

Connective Tissue Contents of Arteries and Taenia Coli[a]

Animal and tissue	No. of experiments	C (% DS)	E (% DS)	C + E (% DS)	ECS/ (C + E)	Hexuronic acid % DS	Hexuronic acid mmol/kg DS	Hexosamine % DS	Hexosamine mmol/kg DS	Sulfate % DS	Sulfate mmol/kg DS	Reference
Dog												
Carotid artery	7	51.4 ±1.4	24.0 ±1.3	75.4 ±2.6	2.1	0.33 ±0.02	17.0 ±1.2	0.39 ±0.01	21.9 ±0.7	—	—	Jones and Swain, 1972
	6	—	—	—	—	—	—	0.84 ±0.06	46.7 ±3.3	0.14 ±0.01	15.0 ±0.9	Villamil and Matloff, 1975
Lingual artery	10	44.6 ±0.9	14.8 ±0.6	59.4 ±1.5	3.2	0.33 ±0.01	17.0 ±0.5	0.41 ±0.01	23.0 ±0.4	—	—	Jones and Swain, 1972
Rat												
Thoracic aorta	4–7	24.0 ±1.0	37.7 ±0.8	61.7 ±1.8	2.1	0.56 ±0.02	28.9 ±1.0	0.43 ±0.01	24.0 ±0.6	—	—	Jones, 1973
Rabbit												
Main pulmonary artery	4–6	17.5 ±0.3	18.9 ±0.8	36.4 ±1.0	7.1	0.54 ±0.02	27.8 ±1.0	0.52 ±0.02	28.9 ±1.0	—	—	Jones, 1980
Guinea pig												
Taenia coli	6	8.9 ±0.7	1.0 ±0.2	9.9 ±0.7	23.6	0.22 ±0.01	11.5 ±0.5	0.35 ±0.01	19.4 ±0.5	—	—	Jones and Swain, 1972

[a] Values are mean ± SEM. ECS, Extracellular space; C, collagen; E, elastin; DS, dry solids.

The extracellular water and connective tissue contents allow arterial electrolyte to be further partitioned. A first approximation is to correct the total electrolyte for that dissolved in ECS, and these values appear in Table III for the tissues whose water distribution is shown in Table I. The estimates of the K concentration in cell water, $[K]_i$, are consistent at about 170 m*M*. The corrections for Na and Cl are large, and the resultant values show a twofold range. The $[Na]_i$ was calculated to be lowest for taenia coli and the $[Cl]_i$ among the lowest. The taenia coli contains the lowest amount of collagen, elastin, and glycosaminoglycans. It appears that further corrections are required to account for ion adsorption to connective tissue.

Cationic adsorption is prevalent in tissues with a high content of glycosaminoglycans (Manery, 1954). Chloride adsorption to collagen has also been reported in the rabbit. Two approaches to the estimates of ionic adsorption appear in Table IV. One involves the measurement of connective tissue contents and ionic adsorption in model systems (Jones and Swain, 1972). Another involves the measure of fast-exchanging electrolyte in excess of the ECS (Jones and Karreman, 1969a; Jones and Swain, 1972; Friedman *et al.*, 1974). The estimate of adsorbed Na based on the fast exchange is slightly greater than that based on connective tissue content, since some of the fast-exchanging electrolyte originates from damaged cells. The more conservative approach is based on the analysis of glycosaminoglycans. For the purpose of estimating ionic adsorption, measurement of hexuronic acid and hexosamine (or hydrolyzable sulfate) should yield a reasonable estimate of the sites for ion adsorption (Jones, 1980; Mathews, 1975; Serafini-Fracassini and Smith, 1974). The affinity for Ca is reported to be slightly greater than for Na (Dunstone, 1962), but at physiological concentrations about 75% of the acid glycosaminoglycans are in the Na form (MacGregor and Bowness, 1971). The Na form would be 30 times more prevalent than the K form given similar selectivities and the adsorption of K can be neglected for most computations. From these considerations, an estimate of adsorbed Na can be derived by multiplying the hexuronic acid plus hexosamine contents (in millimoles per kilogram of dry solids) by 0.75. If only one of the residues was measured, then multiplying its content by 1.5 would yield an equivalent result. Analyses of adventitia and tendon (Table IV) indicate that the adsorption of cation to fibrous connective tissue is negligible.

Correction of the Na contents in Table III for the amount adsorbed to glycosaminoglycans or excess fast-exchanging components led to a significant reduction in the estimated $[Na]_i$. The values of 10–40 mmol/kg cell water were in general agreement with those derived from the

TABLE III

Electrolyte Distribution in Arteries and Taenia Coli[a]

Animal and tissue	No. of experiments	Total (mmol/kg DS)			Total − ECS (mmol/kg DS)			Concentration (mmol/kg cell water)					References
		Na	K	Cl	Na	K	Cl	$[Na]_i$	$[K]_i$	$[Cl]_i$	$[Na]_i^b$	$[Cl]_i^c$	
Dog													
Carotid artery	6	369 ± 17	128 ± 9	351 ± 11	44 ± 13	118 ± 9	31 ± 19	64 ± 18	174 ± 13	45 ± 28	22	—	Cox *et al.*, 1976
Lingual artery	10	332 ± 12	204 ± 5	392 ± 6	49 ± 11	195 ± 5	109 ± 8	40 ± 9	164 ± 7	91 ± 5	16	—	Jones and Swain, 1972
Rat													
Thoracic aorta	8	324 ± 11	134 ± 2	346 ± 12	48 ± 5	114 ± 2	64 ± 5	63 ± 5	151 ± 3	85 ± 7	10	—	Jones and Hart, 1975
Tail artery	60	369 ± 2	224 ± 1	—	74	214	—	58	167	—	23[d]	—	Friedman *et al.*, 1974
	7	394 ± 5	235 ± 5	345 ± 7	98	225	84	76	176	66	42[d]	—	Palatý *et al.*, 1971
Rabbit													
Main pulmonary artery	8	470 ± 9	242 ± 12	479 ± 14	97 ± 8	227 ± 11	158 ± 15	66	154	107	37	101	Jones *et al.*, 1973
Guinea pig													
Taenia coli	9	364 ± 8	510 ± 12	418 ± 8	71 ± 5	497 ± 12	153 ± 6	26	181	56	18	—	Jones *et al.*, 1973

[a] Values are means ± SEM. ECS, Extracellular space; DS, dry solids.
[b] Corrected for adsorption to acid glycosaminoglycans.
[c] Corrected for adsorption to collagen.
[d] Corrected for excess fast-exchanging Na.

TABLE IV

Adsorbed Electrolytes in Arteries, Connective Tissue, and Taenia Coli

Animal and tissue	Method	No. of experiments	Adsorbed content (mmol/kg DS)[a] Na	K	Cl	References
Dog						
Carotid artery	Glycosaminoglycans	7	29	—	—	Jones and Swain, 1972
	Excess fast isotope	6–9	53	5	26	Jones and Karraman, 1969a
Carotid adventitia	Electrolyte analysis	7–22	0	0	0	Jones and Karraman, 1969a
Lingual artery	Glycosaminoglycans	10	30	—	—	Jones and Swain, 1972
	Excess fast isotope	7	32	—	—	Jones and Swain, 1972
Rat						
Thoracic aorta	Glycosaminoglycans	7	40	—	—	Jones, 1973
Tail artery	Excess fast exchange	—	44	—	—	Friedman *et al.*, 1974
Rabbit						
Main pulmonary artery	Glycosaminoglycans	6	42	—	—	Jones[b]
Tendon	Electrolyte analysis	14	0	—	60	Brading and Jones, 1969
Guinea pig						
Taenia coli	Glycosaminoglycans	6	23	—	—	Jones and Swain, 1972

[a] DS, Dry solids.
[b] Unpublished observations.

analysis of isotope effluxes (Table V). The $[Na]_i$ for arterial preparations also showed good agreement with that for taenia coli which had undergone smaller corrections for adsorbed Na. Evidence for Cl adsorption was found only in the rabbit, and the correction was relatively minor. The $[Cl]_i$ values derived by chemical dissection are well above those predicted for a passive distribution (Jones, 1980). Isotope flux and electron probe analyses provide a more detailed test of this observation.

B. Flux Analyses

The application of radioisotope flux analyses allows a more precise measure of cellular electrolyte levels and in addition yields information on kinetic behavior of cellular components. Analyses of isotope fluxes have led to partitioning of electrolyte into fast- and slowly exchanging components. Manipulation of experimental conditions such as tempera-

TABLE V

Turnover and Content of Electrolytes in Arterial Smooth Muscle and Taenia Coli Determined by Radioisotope Methods[a]

			Slow-exchanging electrolyte									
			Content (mmol/kg DS)			Content (mmol/kg cell water)			Turnover (min^{-1})			
Animal and tissue	No. of experiments	Temp. (°C)	Na	K	Cl	$[Na]_i$	$[K]_i$	$[Cl]_i$	Na	K	Cl	References
Dog												
Carotid artery	6–8	37	20	108	31	23	126	36	0.035	0.0045	0.038	Jones and Karraman, 1969a
Lingual artery	7	2	12 ± 1	—	—	10	—	—	0.0021	—	—	Jones and Swain, 1972
	7	37	7	—	—	6	—	—	0.018 ± 0.002	—	—	Jones and Swain, 1972
Rat												
Thoracic aorta	6–8	37	—	119 ± 5	38 ± 1	—	156	48	—	0.0090 ± 0.0002	0.136 ± 0.004	Jones and Hart, 1975; Jones *et al.*, 1977
	4	1–37	9	—	—	12	—	—	0.22 ± 0.02	—	—	Jones, 1974

(continued)

TABLE V (*Continued*)

Animal and tissue	No. of experiments	Temp. (°C)	Slow-exchanging electrolyte: Content (mmol/kg DS) Na	K	Cl	Content (mmol/kg cell water) $[Na]_i$	$[K]_i$	$[Cl]_i$	Turnover (min^{-1}) Na	K	Cl	References
Tail artery	6–8	37	—	—	—	—	—	—	0.16 ± 0.01	0.0064 ± 0.0005	0.102 ± 0.006	Dutta and Jones[d]
	—	35	13	—	—	8	—	—	0.145 ± 0.005	—	—	Garay *et al.*, 1979
	41	37–2	26 ± 1[b]	218 ± 1[b]	—	21	170	—	—	—	—	Friedman *et al.*, 1974
Rabbit												
Main pulmonary artery	7–14	37	—	273 ± 6[c]	123	—	158[c]	47	—	0.0038 ± 0.0002	0.037 ± 0.001	Jones *et al.*, 1973; Miller, 1977
	7	1–37	21 ± 2	—	—	12	—	—	0.095 ± 0.005	—	—	Miller, 1977
Guinea pig												
Taenia coli	10–12	35	34 ± 2	436[c]	154 ± 1	13	164[c]	58	0.24	0.0102	0.065	Casteels, 1969

[a] Values are means ± SEM. DS, Dry solids.
[b] Determined by a lithium-exchange method.
[c] Determined by chemical dissection methods.
[d] Unpublished observations.

ture, membrane active agonists, ionic concentrations, and transport inhibitors allows membrane-limited components to be identified. A detailed discussion of experimental methods and formal mathematical analyses appears elsewhere (Jones, 1975, 1980).

The washout of ^{42}K and ^{36}Cl, shown in Figs. 1 and 2, approaches a single exponential after the initial rapid clearing of the ECS. Extrapolation to time zero yields an estimate of the slow counts which are most likely of cellular origin (Jones, 1980). Identification of cellular Na is made difficult by the large ECS. A washout conducted at 1°C following loading with ^{24}Na at 37°C allows fast- and slowly exchanging components to be separated as shown in Fig. 3. A return to 37°C increases the efflux markedly. This increase in rate can be significantly inhibited by first exposing the tissue to ouabain (Fig. 3). The ouabain-dependent efflux is equated with the active transport of Na. Further evidence that the slow ^{24}Na component is of cellular origin results from the more than 10-fold increase in this component when the tissues are first equilibrated in a K-free solution (active transport-inhibited) (Jones and Karreman, 1969a; Jones and Swain, 1972).

The amounts and rate constants for the slowly exchanging Na, K, and Cl are given in Table V. The cellular Na values (10–20 mmol/kg cell water) are similar to those in Table III, which resulted from chemical dissection. There is less variation with the isotope method, however. The slowly exchanging $[Cl]_i$ is lower than that derived from chemical dissection but exceeds by three- to fourfold the $[Cl]_i$ predicted for a passive distribution (Jones, 1980). Chemical dissection and isotope flux analyses

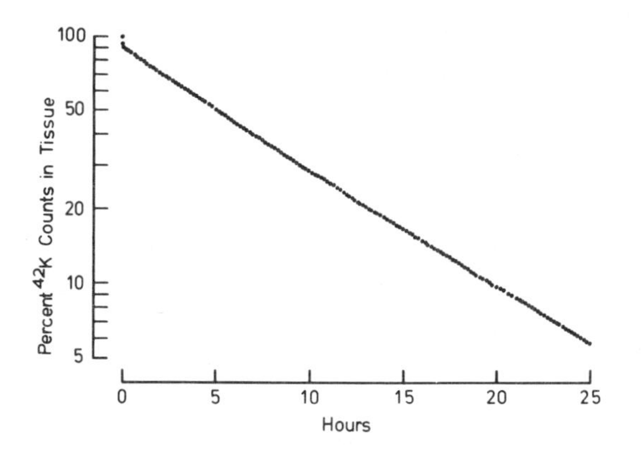

Fig. 1. Steady-state washout of ^{42}K from a dog lingual artery incubated in normal physiological salt solution ($[K]_o$ = 5.0 mM). Log of percent counts is plotted versus time. The rate constant at 25 hr (0.0018 min^{-1}) is within reasonable agreement with that at 1 hr (0.0022 min^{-1}).

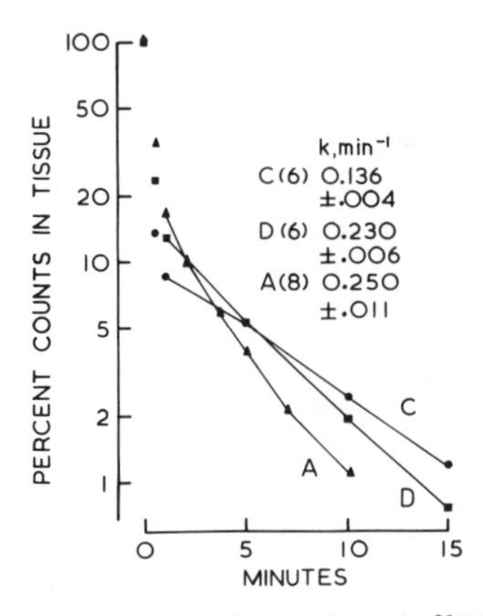

Fig. 2. Representative steady-state washout of aortic ^{36}Cl from control (C), DOCA-hypertensive (D), and aldosterone-hypertensive (A) rats. Data are plotted as in Fig. 1. Average rate constants ± SEM (number of rats) are given with D and A significantly greater than C ($p < .001$).

yield equivalent results for the taenia coli, a preparation which has a relatively small contribution from connective tissue. The $[K]_i$ values were consistent for the two methods and were in agreement with those for other tissues. The $[K]_i$ was similar for the vascular sites studied.

The rate constants showed some species differences. The rate constants for the rat aorta were twice that for dog carotid or rabbit main pulmonary artery (Table V). In the rat the more peripheral arteries had a lower turnover than the aorta. Some differences are seen in ^{24}Na rates, but one must be cautious about different methods for determining the rate constants (Jones, 1980). It is of interest that the ^{24}Na rate is highest in the rat, attaining a value similar to that for the taenia coli. The rapid turnover (half-time ≈ 2–3 min) underscores the difficulties in analyzing this component. The rate constants for ^{36}Cl are intermediate between those for ^{42}K and ^{24}Na, and some show species effects, the rat again showing the highest rate. The cellular Cl behaves largely as a uniform compartment. The question whether the high $[Cl]_i$ is uniformly distributed or is sequestered in cellular organelles (hence free Cl may be low) can be approached by electron probe analyses.

C. Electron Probe Analyses

The electron probe provides a measure of intracellular content and distribution of elements. This technology has been developed and applied to the study of vascular smooth muscle in the laboratory of the Somlyos. A

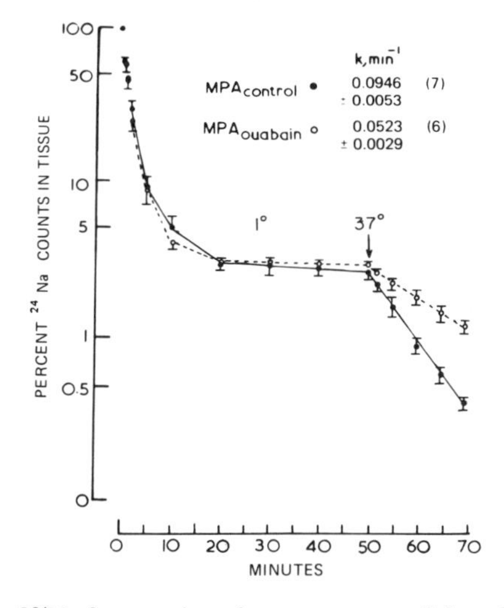

Fig. 3. Washout of ^{24}Na from main pulmonary artery of the rabbit plotted as in Fig. 1. The tissues were loaded in $[K]_o = 5$ m*M*, with the ^{24}Na solution at 37° and at 1°C for the final 5 min. Washouts were initiated at 1°C either in normal physiological solution (control) or with added ouabain (10^{-5}*M*). The tissues were returned to 37°C at 50 min, and rate constants ± SEM were derived from the initial part of the efflux. The difference between the two conditions (ouabain-sensitive rate) is operationally defined to be active transport. (Data from Miller, 1977).

discussion of the method and its limits appears elsewhere (Jones, 1980; Shuman *et al.,* 1976, 1977; Johansson and Somlyo, 1980). Of critical importance to this method is the cryoultramicrotomy. The tissues have to be quickly frozen to stop diffusion of ions and the formation of ice crystals. At this time the best vascular data are derived from the portal vein of the rabbit (a muscular yet thin-walled vessel).

An important feature of the electron probe is the ability to perform elemental analysis on organelles such as mitochondria and nuclei. The results of a recent study (Somlyo *et al.,* 1979) are summarized in Table VI. For comparison isotope and chemical analyses of rabbit portal vein are also presented. These latter analyses assume a uniform cellular distribution. The K, Cl, and Mg levels show good agreement between the probe and other methods. Little evidence is seen for significant gradients between the cytoplasm and the nucleus. The K and Cl levels appear to be slightly lower in the mitochondria, whereas the Mg levels are similar to cytoplasmic values. There is no major evidence for sequestered Cl, and the high concentration results from the action of an energy-requiring transport system.

Sodium appeared to be distributed uniformly throughout the cells,

TABLE VI

Electrolyte Content of Vascular Smooth Muscle and Organelles in Rabbit Portal Vein Determined by Electron Probe, Isotope, and Chemical Analyses

			Content (mmol/kg DS)[a]				Content (mmol/kg cell H_2O)[a]				
Method	Component	No. of animals	K	Na	Cl	Mg	K	Na	Cl	Mg	References
Electron probe	Cytoplasm	9	611	167	278	36	172	47	78	10	Somlyo *et al.*, 1979
	Nucleus	6	649	177	256	39	183	50	72	11	Somlyo *et al.*, 1979
	Mitochondria	9	464	193	220	38	131	54	62	11	Somlyo *et al.*, 1979
Isotope	Whole cell	6–7	—	92	234	—	—	26	66	—	Miller, 1977
Chemical	Whole cell	8–9	649	81	177	41	183	23	50	12	Jones *et al.*, 1973; Jones, 1980

[a] Original data were interconverted between dry solids (DS) and cell water based on 78% cell water (Somlyo *et al.*, 1979).

but unlike the other ions showed probe analysis (Table VI) values about twice those derived from analysis of isotope effluxes. This may be attributed to differences in the initial conditions employed in the two methods (Somlyo *et al.*, 1979). Relatively long incubation times and reduced stretch may contribute to the lower values associated with the isotope efflux analyses. The probe analyses have borne out the assumption that cellular Na, K, and Cl are uniformly distributed. This has important consequences for kinetic analyses of transport systems.

III. ELECTROLYTE TRANSPORT

The ionic concentrations and turnover discussed above result from the action of several transport mechanisms operating across the cell membrane. Some fluxes, e.g., K efflux, are thought to occur mainly via one mechanism (permeation or simple diffusion), while other fluxes involve permeation, exchange diffusion, and active transport simultaneously. Alterations occur in the steady-state turnover of electrolytes during experimental hypertension (Jones, 1982), which further underscores the importance of obtaining a characterization of the individual transport mechanisms. Considerable information is available for some transport processes (permeation), while little is available for others (exchange diffusion) in vascular smooth muscle. It is anticipated that the study of specific transport mechanisms will become an increasingly important part of future approaches to the electrolyte metabolism of the arterial wall.

A. Permeation

The random movement of ions through the membrane plays a major role in the development of a membrane potential. The permeability coefficient is a basic parameter which represents the ease with which an ion diffuses through the membrane. As ions diffuse along their concentration gradients, a diffusional potential E_{diff} is thought to develop according to the Goldman equation:

$$E_{\text{diff}} = \frac{RT}{zF} \ln \frac{P_K[K]_o + P_{Na}[Na]_o + P_{Cl}[Cl]_i}{P_K[K]_i + P_{Na}[Na]_i + P_{Cl}[Cl]_o} \tag{1}$$

where P_K, P_{Na}, and P_{Cl} are the respective permeability coefficients, R the gas constant, T the absolute temperatue, z the valence of the ion, and F the Faraday constant. The relative permeability of the membrane to K, Na, and Cl is an important determinant of the diffusional potential. The permeability coefficients can be estimated experimentally from the measure of passive fluxes and the prevailing electrochemical gradients, which are summarized in Table VII. The details of the computations are presented in Jones (1980). The concentration gradients and membrane potential (E_m) are similar for the three preparations shown. The passive fluxes represent the major difference among tissues. The total ^{42}K and ^{36}Cl effluxes were taken to occur via passive diffusion or permeation. This assumption may not be entirely correct, especially for Cl, but at present little information is available which supports the presence of alternative transport mechanisms. The passive Na influx was estimated in vascular smooth muscle from the active transport of Na (K-dependent or ouabain-sensitive efflux). In the pump-leak model, active transport of Na out of the cell is balanced by passive diffusion into the cell. The fluxes were computed from the relation

$$\text{Flux} = \text{rate} \times \text{concentration} \times V/A \tag{2}$$

where V/A is the ratio of cellular volume to surface membrane area. This is not easy to quantitate, but a detailed study was conducted on the smooth muscle of the vas deferens (Merrillees, 1968). The average value of 0.6×10^{-4} cm was assumed to be applicable to the preparations in Table VII. The passive flux of Cl exceeded the passive cationic fluxes, while the Na and K fluxes were equivalent.

Smooth muscle membranes are at least five times more permeant to K than to Na. The permeability of Cl varied from about one-half to two times P_K. This may reflect some uncertainty in the estimate of passive Cl

TABLE VII

Membrane Parameters and Diffusion Potentials for Arterial Smooth Muscle and Taenia Coli

	Rat aorta			Rabbit main pulmonary artery			Guinea pig taenia coli		
Membrane function[a]	Na	K	Cl	Na	K	Cl	Na	K	Cl
Cellular concentration (mmol/kg water)	12	156	48	12	158	47	13	164	58
Extracellular concentration (mM)	150	5	145	146	5	141	137	6	134
Turnover (min^{-1})	0.22	0.0090	0.136	0.095	0.0038	0.037	0.24	0.0102	0.065
K- or ouabain-dependent (min^{-1})	0.16	—	—	0.047	—	—	—	—	—
Passive flux (pmol cm^{-2} sec^{-1})[b]	1.9	1.4	6.5	0.6	0.6	1.7	3.1	2.2[c]	3.8
Correction factor[d]	2.37	0.30	2.37	2.50	0.27	2.50	2.37	0.30	2.37
Permeability (cm/sec $\times 10^{-8}$)	0.53	3.0	5.7	0.15	1.4	1.5	1.0	4.5	2.8
E_i (mV)	+67	−92	−30	+67	−92	−29	+62	−89	−22
E_{diff} (mV)	—	−34	—	—	−40	—	—	−33	—
E_m (mV)	—	−55	—	—	−60	—	—	−55	—
$E_p = E_m - E_{diff}$ (mV)	—	−21	—	—	−20	—	—	−22	—
References	Hermsmeyer, 1976; Jones, 1974, 1976			Jones and Miller, 1978; Miller, 1977; Somlyo *et al.*, 1969			Casteels, 1969		

[a] E_i, Equilibrium potential of the ion; E_{diff}, diffusion potential; E_m, membrane potential; E_p, electrogenic potential.
[b] Original data were all converted to $V/A = 0.6 \times 10^{-4}$ cm.
[c] Includes correction for diffusion delay.
[d] Factor for electrical gradient (see Jones, 1980, Eqs. 12 and 13).

effluxes. Unlike those for many excitable tissues, the equilibrium potentials for K and for Cl differ greatly from E_m. The highly positive E_{Na} is consistent with that for other excitable tissues. The discrepancy between E_{Cl} and E_m indicates a nonpassive distribution such that Cl is actively transported into smooth muscle. The E_{diff} computed from the data in Table I and Eq. (1) are significantly less electronegative than E_m by about 20 mV. This difference has been attributed to an electrogenic pump (E_p) (Anderson, 1976; Casteels, 1969), but it has been questioned whether differences of more than 10 mV can be maintained during a steady state by such a mechanism (Brading and Widdicombe, 1974). Other sources for the difference between E_m and E_{diff} may result from inadequate estimates of the permeabilities and intracellular concentrations. There is still a need to provide a quantitative basis for the resting membrane potential.

Some general conclusions appear warranted despite the uncertainties. The P_K is similar to P_{Cl} and in turn is 5- to 10-fold greater than P_{Na}. The P_K and P_{Cl} are about 10 times less than the corresponding permeabilities in skeletal muscle (Hodgkin and Horowicz, 1959). The vascular smooth muscle membrane therefore is a less conductive membrane. High $[Cl]_i$ constitutes a major difference in ionic gradients, which cannot be attributed to selected accumulation by cellular organelles. Direct measures of Cl activity would be helpful in clarifying this point.

B. Active Transport

The transport of Na and K against an electrochemical gradient is an energy-requiring process which utilizes ATP as a metabolic substrate. As discussed below, inhibition of the energy metabolism results in a loss of ionic gradients. Also, gradients are removed as a result of selective blockage with cardiac glycosides. A Mg-ATPase has been identified in membrane fragments from arteries, which shares properties with the transport mechanism (Allen and Bukoski, this volume, Chapter 4; Allen and Seidel, 1977; Wei *et al.,* 1976). For instance, the enzyme is sensitive to similar concentrations of ouabain. The transport ATPase is only a small percentage of the total ATPase in the isolated fractions, which makes detailed characterization difficult. It would be helpful to have quantitative information about the K- and Na-stimulating properties in order to link activation of the enzyme with the ouabain-sensitive fluxes.

Part of our recent work has focused on the kinetics of active Na and K fluxes in arterial smooth muscle (Jones, 1981b; Heidlage and Jones, 1981) and has been recently reviewed (Jones, 1980). The protocol shown in Fig. 3 was used to identify ouabain-sensitive or K-dependent ^{24}Na

effluxes which were operationally defined to be active transport. The washouts were conducted in the presence of varying $[K]_o$ or $[Na]_i$ to derive the respective stimulation curve. The active influx of ^{42}K was determined from the 10-min uptake of ^{42}K from solutions of varying $[K]_o$ with and without ouabain. A ouabain-inhibited influx of Rb into Na-loaded arteries had previously been reported (Overbeck *et al.*, 1976), but a systematic evaluation of the stimulation curves had not been made.

The ouabain-dependent ^{24}Na effluxes and ^{42}K influxes derived from rabbit carotid arteries appear in Fig. 4. These were determined in tissues

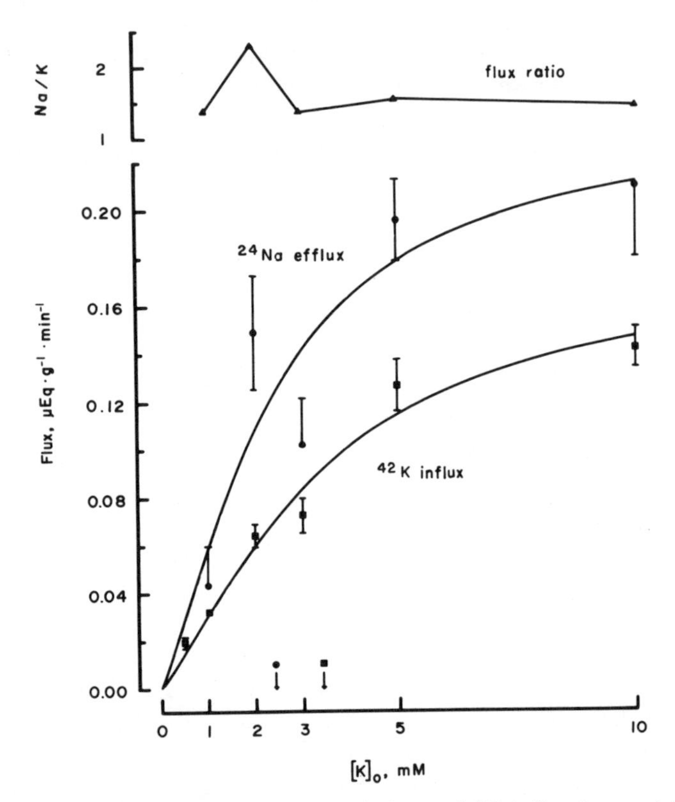

Fig. 4. Ouabain-dependent ^{24}Na efflux (circles) and ^{42}K influx (squares) for rabbit carotid artery at various $[K]_o$. Averages ± SEM from 8–12 rabbits are plotted. The fluxes can be converted to a kilogram cell water basis by multiplying by a factor of 6.0 based on 16.7% of the wet weight that is cell water. The ratios of active Na efflux to K influx (triangles) are plotted in the upper section. Note that the ratio is greater than 1. Curves were computed from Eq. (3) with the parameters in Table VIII. Arrows indicate the $[K]_o$ required for half-saturation. (By permission from Heidlage and Jones, 1981).

equilibrated under physiological conditions ($[Na]_i$ = 24 and $[K]_i$ = 169 mmol/kg cell water). Several important features are apparent from the activation curves in Fig. 4. Saturation behavior is seen for both ions, indicating the operation of a limited number of transport sites. A sigmoidal curve describes the $[K]_o$ stimulation of active Na and K fluxes (see kinetic model below). This indicates that a multisite transport mechanism is operative. The half-maximal stimulation was achieved at $[K]_o^{0.5}$ = 2.4 and 3.4 m*M*, which is below the physiological $[K]_o$ of 4.5 m*M* in the rabbit. A significant transport reserve is available over the physiological range, however, since the system operates at $\frac{2}{3}T_{max}$ for $[K]_o$ = 5 m*M*. An increase in $[K]_o$ in the environment of the vessel wall *in vivo* would be projected to stimulate active transport.

An especially important finding is a Na/K flux ratio greater than 1 (approximately 1.5) over the range studied. This provides a chemical basis for the observation that increased $[K]_o$ over physiological levels has a hyperpolarizing effect on E_m (Anderson, 1976; Hendrickx and Casteels, 1974; Hermsmeyer, 1976). Such hyperpolarization is associated with reduced contractile activity (Webb and Bohr, 1979). The data on transport kinetics complement the electrophysiological and mechanical observations and indicate an important role for the electrogenic transport of Na or K in the control of vascular smooth muscle excitability. Under physiological conditions a 5- to 10-mV contribution results from the steady-state operation of active transport (Jones, 1980).

When arteries are Na-loaded, the electrogenic component can exceed −50 mV (Hendrickx and Casteels, 1974). Therefore, in addition to the effect of $[K]_o$, increased $[Na]_i$ has a large stimulatory effect. The dependence of an ouabain-sensitive Na efflux on $[Na]_i$ was studied in rabbit carotid artery, while a K-dependent efflux was measured in rat aorta (Fig. 5). The rat is relatively insensitive to ouabain, but 2 m*M* was observed to reduce Na efflux to levels achieved in a K-free solution (Jones, 1981b). Intermediate $[Na]_i$ values were obtained by equilibrating the tissues with low $[K]_o$ (e.g., 1.0, 1.5 m*M*) during the ^{24}Na loading. The active effluxes showed a sigmoidal dependence on $[Na]_i$ and saturation behavior. The maximum efflux was greater in the rat than in the rabbit, which confirms the observation made under steady-state conditions (Table V). Half-saturation was achieved at $[Na]_i^{0.5} \approx$ 50 mmol/kg cell water for the rabbit, while the relation was shifted to the left in the rat ($[Na]_i^{0.5} \approx$ 20 mmol/kg cell water). These values were about two times the respective $[Na]_i$ at which each tissue operated in these experiments (Fig. 5). Therefore a significant transport reserve exists for both preparations, since the active transport mechanism operates at about 25% of maximum capac-

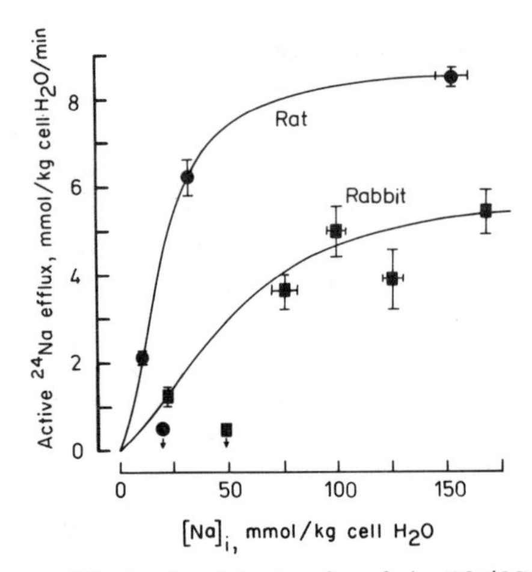

Fig. 5. Dependence of K-stimulated (rat) and ouabain ($10^{-4}M$)-sensitive (rabbit) ^{24}Na efflux on $[Na]_i$. The various $[Na]_i$ values were achieved by loading the tissues in solutions of varying $[K]_o$ (zero for highest $[Na]_i$, 1–1.5 mM for intermediate $[Na]_i$, and 5 mM for low $[Na]_i$). A $[K]_o$ of 10 mM was used for the washout. Averages ± SEM for efflux and $[Na]_i$ are presented for six to eight animals each. Curves were computed from Eq. (3) with the parameters in Table VIII. Arrows indicate the $[Na]_i$ required for half-saturation. Note that the curve for the rabbit carotid is shifted to the right. (Replotted from data in Jones, 1981b; and Heidlage and Jones, 1981).

ity. Comparison to the $[K]_o$ stimulation curves indicates that an increase in $[Na]_i$ would be a more effective activator of the pump, but combinations of the two ions would be needed for maximum effects. The net effect of the sigmoidal kinetics is to make vascular smooth muscle quite responsive to changes in $[Na]_i$, which results in a high gain and more efficient regulation over the physiological range than would be expected from Michaelis–Menten kinetics.

Kinetic analyses of the data in Figs. 4 and 5 were performed with a transport model based on cooperative interaction between multiple transport sites (Jones, 1980). Several other models which predict sigmoidal behavior could fit the data equally well (Jones, 1980). They tend to take the form of polynomials. The cooperative interaction model appearing in Eq. (3) was applied because (1) earlier work from this laboratory supported the concept of cooperative interaction between ions at adsorption sites (Jones and Karreman, 1969b), (2) a minimal number of parameters are optimized (n is the parameter of nearest-neighbor interaction, κ the selectivity coefficient, and T_{max} the maximum transport flux), and (3) the model was easily fitted by a nonlinear least squares fit (derived by Francis Heidlage). The kinetic model is

$$T = \frac{T_{\max}}{2}\left(1 + \left[\frac{\left(\frac{[\mathrm{K}]_o}{[\mathrm{Na}]_o}\kappa_o\right) - 1}{\left\{\left(\frac{[\mathrm{K}]_o}{[\mathrm{Na}]_o}\kappa_o - 1\right)^2 + 4\frac{[\mathrm{K}]_o}{[\mathrm{Na}]_o}\kappa_o n^{-2}\right\}^{1/2}}\right]\right) \tag{3}$$

For $n > 1$ an autocooperative interaction (self-reinforcing) occurs between transport sites. That is, the adsorption of K at an outside site reinforces the adsorption of K at a nearest-neighbor site, which results in a sigmoidal relation between $[\mathrm{K}]_o$ and active transport of Na T^{Na} and K_i T^{K}. A similar cooperative interaction takes place on the inside of the membrane, with the adsorption of Na reinforcing the further adsorption of Na at a nearest-neighbor site. This approach emphasizes the adsorption of ions on the transport molecule(s) as an important determinant of the membrane active flux. Under conditions where the analysis shows $n = 1$, according to this approach no interaction occurs and Eq. (3) reduces to Michaelis–Menten kinetics.

The parameters for active transport in rabbit carotid artery and rat aorta appear in Table VIII. These were derived from the least squares fit routine. When $[\mathrm{K}]_o$ was varied, the sites on the outside of the membrane showed much greater selectivity (κ_o) for K over Na. Varying $[\mathrm{Na}]_i$ revealed that the selectivity of the inside sites (κ_i) was greater for Na than for K, but the absolute value was about 1/10 that exhibited by the outside sites. The κ_o values derived from similar protocols were similar for the rat and rabbit preparations. On the other hand, the rat exhibited about two times greater inside selectivity (κ_i) than the rabbit. This may be related to the differences observed in the steady-state $[\mathrm{Na}]_i$ in rabbit and rat arteries (Table VIII). The parameter for cooperative interaction was greater than 1 in all cases. The values for inside and outside activation curves exhibited good agreement for the rabbit carotid, while the rat had a much higher value for the inside sites. The rat aorta had a higher $T_{\max}$ than the rabbit aorta. Further experiments are needed to test whether this is the result of different cell geometry, number of transport sites, or turnover at the individual sites. In general the kinetics for active transport in vascular smooth muscle are similar to those of other cell types such as red blood cells (Hoffman and Tosteson, 1971). The major difference is $T_{\max}$, which may reflect differences in the number of transport sites more than inherent differences in the properties of the transport proteins. A test of this conclusion must await extraction, purification, and characterization of the active transport complex from vascular smooth muscle, however.

TABLE VIII

Kinetic Parameters for the Active Transport of Na and K by Arteries

Animal and tissue	No. of experiments	$[K]_o$ varied at physiological $[Na]_o$						Reference
		$[Na]_i$ (mmol/kg water)	T_{max}^{Na} (mmol kg^{-1} water min^{-1})	T_{max}^{K} (mmol kg^{-1} water min^{-1})	$[K]_o^{0.5}$ (mM)	κ_o[a]	n	
Rat, aorta	5–8	11	2.5	—	2.3	65	1.1	Jones, 1981b
Rabbit, carotid artery	6–12	24	1.5	—	2.4	66	1.4	Heidlage and Jones, 1981
	6–12	24	—	1.1	3.4	46	1.4	

Animal and tissue	No. of experiments	$[Na]_i$ varied at $[K]_o = 10mM$				Reference
		T_{max}^{Na} (mmol kg^{-1} water min^{-1})	$[Na]_i^{0.5}$ (mmol/kg water)	κ_i[b]	n	
Rat, aorta	6–8	8.5	20	7.0	1.8	Jones, 1981b
Rabbit, carotid artery	6	5.3	49	3.0	1.2	Heidlage and Jones, 1981

[a] Selectivity coefficient of outside to K_o over Na_o.
[b] Selectivity coefficient of inside to Na_i over K_i.

C. Exchange Diffusion

Transport processes which move ions across the membrane in a mole-for-mole exchange for a similar ion have received little study with vascular preparations. Exchange diffusion is only indirectly coupled with energy metabolism and more directly derives its driving forces from the prevailing ionic concentration gradients and kinetic energy of the molecules. A one-for-one exchange of Na was identified when active transport was blocked in taenia coli of the guinea pig (Brading, 1975). The residual flux depended on the availabilityof Na on the opposite side of the membrane. As $[Na]_o$ was raised, the Na efflux followed Michaelis–Menten kinetics. Increased $[Na]_i$ caused the Na efflux to reach a maximum. This response was also observed with the ouabain-insensitive efflux of ^{24}Na from rabbit carotid artery as $[Na]_i$ was varied (Heidlage and Jones, 1981). The dependence of this efflux on $[Na]_o$ and competitive ions like Li has not been studied in arteries. There is one report of ^{36}Cl efflux from dog carotid artery being dependent on $[Cl]_o$ (Villamil *et al.*, 1968b). No such dependence of ^{36}Cl efflux was observed, however, on taenia coli of the guinea pig (Casteels, 1971).

Blaustein (1977) has proposed that a variant of exchange diffusion operates to regulate the transmembrane levels of Ca. According to this model, Na and Ca compete for sites in a countertransport system. Under physiological conditions the Na gradient is proposed to move Na into the cell in exchange for Ca. Factors which elevate $[Na]_i$, e.g., inhibition of active transport and increased leakage of Na during hypertension, would reduce the efflux of Ca, leading to increased contraction. Evidence for and against this mechanism has been reviewed (Jones, 1980; Van Breemen *et al.*, 1980) but, as in the case of other exchange diffusion processes, there is need for data to provide a definitive test of the model from which kinetic parameters can be derived.

IV. METABOLIC CONTROL

Sufficient energy must be produced by the cell to meet the requirements of the electrochemical work performed by transport processes. It is therefore expected that a close link exists between the energy supply and the ability to maintain ionic gradients. It is important to establish the energetic feasibility of the active transport model proposed above, since no model, no matter how well it fits the data, can violate fundamental laws of thermodynamics. Once established, it is of interest to follow the responses of the transport systems to metabolic inhibition during reduced temperature and metabolic activity achieved by selective inhibition of aerobic and anaerobic metabolism.

A. Transport Work and Efficiency

Data are available which allow a comparison in rat aorta between the required transport power and the available energy. Transport power can be computed under steady-state conditions from the product of active flux and the electrochemical gradient (Table VII). Production of the energy substrate, ATP, can be estimated from steady-state oxygen consumption (Q_{O_2}). Two studies (Briggs *et al.*, 1949; Krantz *et al.*, 1951) indicate that a Q_{O_2} of 1.03 μl/mg dry solids per hour is a representative value with glucose as a substrate. This is equivalent to 6 mmol ATP/kg cell water per minute (see Jones, 1980, for details). Based on the estimation that 10,000 cal is available per mole ATP hydrolyzed under physiological conditions (Lehninger, 1971), the metabolic power in rat aorta is 60 cal/kg cell water per minute.

The power required to transport Na against the electrical (−55 mV) and chemical (12/155) gradients is −5.4 cal/kg cell water per minute or 9% of the ATP power (see Jones, 1980, for details). A similar computation for active K influx, assuming a 3 Na/2 K ratio, yields −1.1 cal/kg cell water per minute or 1.8% of the ATP power. Therefore, about 11% of the total ATP power is required to support the active transport of Na and K in rat aortic smooth muscle. This mechanism is therefore energetically feasible even at efficiencies under 50%. The energy requirements for other transport systems have also been estimated (Jones, 1980). Active Cl influx requires −2.3 cal/kg cell water per minute, and Ca efflux −1.6 cal/kg cell water per minute. The total transport power required for the four ions is −10.4 cal/kg cell water per minute or 17.3% of the total ATP power. At 50% efficiency, the transport of ions would consume about one-third of the resting energy production, which is feasible.

It has been proposed that 1 ATP is consumed per complete cycle of the pump in some tissues (Thomas, 1972). For vascular smooth muscle, 3 Na plus 2 K would require about 10,200 cal per cycle (Jones, 1980). This is equivalent to 1 ATP at 100% efficiency. Two ATP are required per cycle (or 1.5 Na and 1 K per ATP) for a pump operating at 50% efficiency. The ratio of 2 ATP per cycle is similar to that derived from studies on active transport in taenia coli maintained under anaerobic conditions (Casteels and Wuytack, 1975). It therefore appears that the active transport mechanism in vascular smooth muscle requires 2 ATP per cycle of 3 Na and 2 K.

B. Temperature Dependence

Changes in temperature have often been employed in studying mechanisms of membrane transport. Processes that proceed along established electrochemical gradients, such as simple diffusion, should show only a small temperature dependence, while processes that involve enzymatic activity, such as active transport, should exhibit a large temperature dependence. Arrhenius derived the relation between temperature and rate constants (k), which takes the form

$$\log k = -(E_a/2.3R)(1/T) + \log \bar{A} \quad (4)$$

where E_a is the activation energy (in calories per mole) and $\bar{A}$ is a constant related to the number of collisions per unit time. A uniform process yields a straight line in an Arrhenius plot. The slope yields E_a. A nonuniform reaction will exhibit a shift from one slope to another often at a discrete transition temperature.

In general, the Na, K, and Cl fluxes in vascular preparations exhibit a high E_a and a related Q_{10} (the change in rate for a 10°C change in temperatue). The total ^{24}Na and ^{36}Cl effluxes from canine carotid have a Q_{10} of 3.0 and 2.6, respectively (Garrahan *et al.*, 1965; Villamil *et al.*, 1968b). Even though ^{36}Cl efflux occurred along an electrochemical gradient, the Q_{10} was about two times that expected for simple diffusion (1.3 or an E_a of 5000 cal/mol). The high Q_{10} for ^{24}Na was confirmed in the portal vein of the rat (Jonsson *et al.*, 1975). Surprisingly, this study showed a high Q_{10} (2.4) and E_a (16,000 cal/mol) for [^{14}C]urea, a compound which permeates the membrane readily. Apparently, the passage of ionic and nonionic solutes across the vascular smooth muscle membrane involves significant interaction with the membrane, even though the overall driving force is along the existing electrochemical gradient.

Although a high E_a does not identify a specific transport mechanism, temperature studies can characterize ion–membrane interactions. Kamm and co-workers (1979b) have made a systematic analysis of the temperature dependence of the active and passive effluxes of ^{24}Na from the rat aorta, as well as the ^{42}K efflux. The data are replotted in Fig. 6. The three effluxes exhibit qualitatively different Arrhenius plots. The K-independent ^{24}Na efflux (mostly exchange diffusion and leak components) followed a linear relation over the entire range studied with a Q_{10} of 3.7 and an E_a of 18,000 cal/mol. The active efflux of Na had two linear ranges with a transition temperature of 17°C. The E_a was increased threefold to 36,000 cal/mol ($Q_{10} = 8.8$) for the low temperatures. This is consistent with the behavior of membrane-bound enzymes

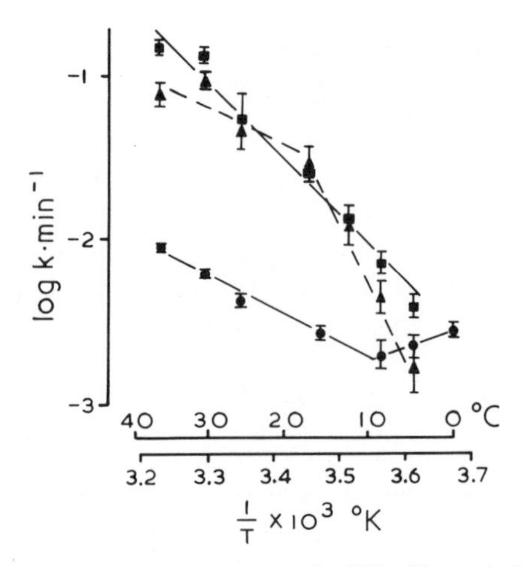

Fig. 6. Arrhenius plot of log rate constants for ^{42}K efflux (circles), K-dependent ^{24}Na efflux (triangles), and ^{24}Na efflux into K-free solution (squares) from 4–12 rat aortas ± SEM. The abscissa is absolute T^{-1}, and the equivalent temperature (°C) is portrayed for comparison. Straight lines were fitted by regression techniques. (From Jones, 1980; adapted from Kamm *et al.,* 1979b).

(Charnock *et al.,* 1973; Raison *et al.,* 1971). The ^{42}K efflux followed a uniform process from 37° to about 10°C (E_a = 10,000 cal/mol). Below this temperature, the rate increased. It has been proposed that the departure from uniformity results from physical changes in the lipids of a membrane which behaves as a fluid mosaic (Kamm *et al.,* 1979b; Nicholson, 1976). For a membrane of a given composition, there is a tendency for a transition from a fluid to a solid state at lower temperatures. Such transitions are often projected to influence the function of membrane transport proteins. The data in Fig. 6 indicate that, if such transitions occur, they are in the immediate domain of the transport sites, since three different relations are seen.

The overall effect of reduced temperature is to slow active transport disproportionately to the reduced leakage of K, especially at temperatures below 15°C. The net effect is a gain in cell Na and a loss of K with a half-time of about 11 hr (Kamm *et al.,* 1979a). In addition to the loss of ionic gradients, smooth muscle cells swell (see Jones, 1980 for a more detailed discussion of cell volume regulation). When the tissues are returned to 37°C, these processes are reversed (Barr *et al.,* 1962; Kamm *et al.,* 1979a). The time course for recovery of canine carotid arteries after 15 hr at 1°C has been measured in our laboratory and is presented in Fig. 7. Rewarming stimulated active transport which resulted in a rapid loss of Na, Cl, and water and an exchange of Na for K. About 60% of the

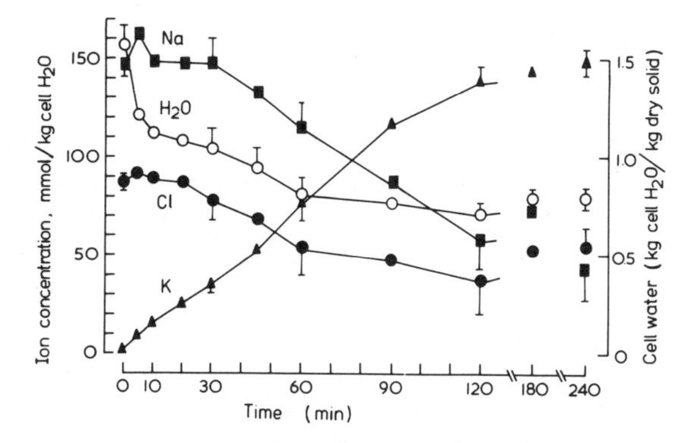

Fig. 7. Time course for recovery of canine carotid arteries at 37°C after cold storage. Averages ± SEM for 11 dogs are plotted. Cell water (right ordinate) was calculated as the difference between total water and the sucrose space. Cell ionic concentrations were calculated as the difference between the total and that contained in the sucrose space and were normalized in terms of cell water. No corrections were made for ion adsorption to connective tissue. The points are joined by straight lines.

total reduction in cell water took place over the first 10 min. The $[Na]_i$ and $[Cl]_i$ were unchanged over this time, which indicates that approximately isosmotic amounts of Na, Cl, and water left the cells. The $[K]_i$ rose linearly over the first 90 min (1.3 mmol/kg cell water per minute). Steady-state levels were achieved after 2 hr. The overall exchange of ions (on a dry or wet basis) can be calculated from $\Delta[Na] - \Delta[Cl] + \Delta[K]$ and is not significantly different from zero [−200 −(−94) + 116 mmol/kg dry solids]. About 45% of the net transport of Na is associated with Cl and cell water, whereas the remainder is coupled with K uptake. ^{42}K influxes and effluxes from rabbit carotid arteries were measured during the initial period of active transport after K-depleted and Na-loaded tissues were returned to K-containing solutions (Heidlage and Jones, 1981). The turnover was over 25 times greater than that during the steady state. The combination of maximal active transport of Na and inability to retain K taken up into the cell may underlie the initial rapid loss of cell water shown in Fig. 7. As the membrane permeability stabilizes at lower levels, more of the K actively transported into the cell is retained, thus adding to development of the $[K]_i$ but reducing the loss of cell water.

C. Metabolic Inhibitors

Vascular smooth muscle sustains a substantial level of aerobic glycolysis, a feature shared with some other smooth muscles and a few other cell types (Paul, 1980). Although oxidative metabolism produces most of the ATP (about 75%) under steady-state conditions, the well-developed glycolytic capability has the potential to provide metabolic support for ion transport under anoxic conditions. This has relevance for local regulation of blood flow to anoxic tissues. It was noted above that about one-third of the oxidative metabolism is required to sustain the steady-state transport of Na, K, Cl, and Ca. Therefore aerobic glycolysis would not be sufficient to support the normal steady state. A substantial increase in lactic acid production would have to take place during anaerobic conditions to maintain normal ion transport. The data indicate that significant increases occur but generally not enough to replace the ATP produced by oxidative metabolism.

A series of experiments have been conducted to test the ability of vascular smooth muscle to accumulate K and to extrude Na in the presence of selective metabolic blockage. Canine carotid arteries were treated similarly to those in Fig. 7. The 32 segments were incubated 15 hr at 1°C in K and glucose-free Krebs solution. Metabolic inhibitors were applied and maintained during the subsequent 3-hr incubation at 37°C [iodoacetate (IAA) was first applied during the last ½ hr in the cold]. The concentrations of the inhibitors and the factorial design appear in Table IX. A complete factorial design was employed for first factors at two levels for 2^5 or 32 combinations. Average K contents (corrected for the $[K]_o$ in the sucrose space during physiological conditions) are shown in Table X. Analyses of variance of the two-way experiment were conducted accord-

TABLE IX

Factorial Design for the Effects of Metabolic Inhibitors on Dog Carotid Artery[a]

Treatment	Without DNP		With DNP (b)	
	Without IAA	With IAA (a)	Without IAA	With IAA (a)
O_2 + Glu − CN	1	a	b	ab
O_2 + Glu + CN (c)	c	ac	bc	abc
O_2 − Glu (d) − CN	d	ad	bd	abd
O_2 − Glu (d) + CN (c)	cd	acd	bcd	abcd
N_2 (e) + Glu − CN	e	ae	be	abe
N_2 (e) + Glu + CN (c)	ce	ace	bce	abce
N_2 (e) − Glu (d) − CN	de	ade	bde	abde
N_2 (e) − Glu (d) + CN (c)	cde	acde	bcde	abcde

[a] O_2, Oxygen (95%); N_2, nitrogen (95%); Glu, glucose (5.5 m*M*); CN, cyanide (4.5 m*M*); IAA, iodoacetate (1 m*M*); DNP, 2,4 dinitrophenol (0.2 m*M*).

TABLE X

Factorial Experiment on the Effects of Metabolic Inhibitors on K Content of Dog Carotid Artery[a]

Treatment	Without DNP		With DNP (b)	
	Without IAA	With IAA (a)	Without IAA	With IAA (a)
O_2 + Glu − CN	130[b]	91	88	70
O_2 + Glu + CN (c)	92	39	70	32
O_2 − Glu (d) − CN	98	65	27	25
O_2 − Glu (d) + CN (c)	15	17	13	17
N_2 (e) + Glu − CN	111	82	76	46
N_2 (e) + Glu + CN (c)	83	37	85	24
N_2 (e) − Glu (d) − CN	20	18	20	14
N_2 (e) − Glu (d) + CN (c)	22	18	13	16

[a] Values represent data for four dogs. K content shown in millimoles per kilogram dry solids (DS). Common standard error of mean ± 2.8 mmol/kg DS.

[b] Contents corrected for 10 mmol K_o/kg DS in sucrose space.

ing to Snedecor and Cochran (1967). The common standard error was estimated from the variance associated with the experimental error (after correction for treatment and animal variation). When used alone, all the metabolic inhibitors reduced the K content ($p < .001$). The coefficients for the main effects and the interactions were computed by the Yates method (Davies, 1960) and appear in Table XI. Student's t test was used to determine the confidence limits, since the 31 main effects and interactions are associated with 31 degrees of freedom.

The main effect of each inhibitor was to decrease the K content of the slices ($p < .001$), but no single inhibitor blocked K accumulation completely. A number of interactions were also apparent. The interaction between two sets of aerobic inhibitors (cyanide and 2,4-dinitrophenol; cyanide and nitrogen) were positive and significant ($p < .001$), while the positive interaction between nitrogen and 2,4-dinitrophenol was suggestive ($.01 < p < .05$). The positive sign indicates that these three inhibitors had similar actions. This would be expected if they acted upon the same metabolic pathway (oxidative phosphorylation). A positive interaction between substrate removal (−Glu) and IAA was observed ($p < .001$), indicating a similar metabolic effect (glycolysis). No significant interactions were evident between glucose removal or IAA and the individual aerobic inhibitors, thereby reaffirming that they acted through different pathways and that the primary effects are additive.

It therefore appears that aerobic oxidative and glycolytic metabolism

TABLE XI

Primary Effects and Interactions of Metabolic Inhibitors on K Content of Dog Carotid Artery[a]

Treatment	Without DNP		With DNP (b)	
	Without IAA	With IAA (a)	Without IAA	With IAA (a)
O_2 + Glu − CN	49.2[b]	−21.9‡	−18.9‡	3.6
O_2 + Glu + CN (c)	−24.3‡	− 2.0	12.4‡	−2.2
O_2 − Glu (d) − CN	−46.2‡	17.2‡	2.9	0.9
O_2 − Glu (d) + CN (c)	5.0	8.1†	0.4	0.1
N_2 (e) + Glu − CN	−13.0‡	0.1	6.8*	−5.0
N_2 (e) + Glu + CN (c)	13.2‡	− 3.1	− 5.7*	1.9
N_2 (e) − Glu (d) − CN	− 4.2	2.3	5.5	1.4
N_2 (e) − Glu (d) + CN (c)	5.6*	− 1.3	− 8.9†	3.1

[a] Values represent data for four dogs. K content shown in millimoles per kilogram dry solids.
[b] Population average.
*$.01 < p < .05$.
†$.001 < p < .01$.
‡$p < .001$.

is essential for vascular smooth muscle to develop normal concentration gradients. The blockage of glycolysis with IAA and/or glucose removal is expected to have some effect on aerobic metabolism by removal of some substrate required for the Krebs cycle. Experiments have not been conducted with substrates other than glucose, In theory, if appropriately supported during inhibition of glycolysis, aerobic metabolism should be sufficient to maintain ionic gradients, but this has yet to be tested. On the other hand, the results of aerobic inhibition show that anaerobic glycolysis is not sufficient to maintain a full level of ion transport, as evidenced by reduced K uptake. This is likely to be related to the inability of vascular smooth muscle to maintain normal ATP levels under these conditions (Paul, 1980).

ACKNOWLEDGMENTS

The authors would like to thank Heather Pace for able assistance. Work on this chapter was partially supported by Public Health Service grant HL 15852.

REFERENCES

Allen, C. J., and Seidel, C. L. (1977). EGTA-stimulated and ouabain-inhibited ATPase of smooth muscle. *In* "Excitation–Contraction Coupling in Smooth Muscle" (R. Casteels, T. Godfraind, and J. C. Rüegg, eds.), pp. 211–218. Elsevier, Amsterdam.

Anderson, D. K. (1976). Cell potential and the sodium-potassium pump in vascular smooth muscle. *Fed. Proc., Fed. Am. Soc. Exp. Biol.* **35,** 1294-1297.

Barr, L., Headings, V. E., and Bohr, D. F. (1962). Potassium and the recovery of arterial smooth muscle after cold storage. *J. Gen. Physiol.* **46,** 19-33.

Blaustein, M. P. (1977). Sodium ions, calcium ions, blood pressure regulation, and hypertension: A reassessment and a hypothesis. *Am. J. Physiol.* **232,** C165-C173.

Bohr, D. F., and Uchida, E. (1967). Individualities of vascular smooth muscle in response to angiotensin. *Circ. Res.* **21,** 135-145.

Brading, A. F. (1975). Sodium/sodium exchange in the smooth muscle of the guinea-pig taenia coli. *J. Physiol. (London)* **251,** 79-105.

Brading, A. F., and Jones, A. W. (1969). Distribution and kinetics of Co-EDTA in smooth muscle and its use as an extracellular marker. *J. Physiol. (London)* **200,** 387-401.

Brading, A. F., and Widdicombe, J. H. (1974). An estimate of sodium/potassium pump activity and the number of pump sites in the smooth muscle of the guinea-pig taenia coli, using [^{3}H]ouabain. *J. Physiol. (London)* **238,** 235-249.

Briggs, F. N., Chernick, S., and Chaikoff, I. L. (1949). The metabolism of arterial tissue. I. Respiration of rat thoracic aorta. *J. Biol. Chem.* **179,** 103-111.

Casteels, R. (1969). Calculation of the membrane potential in smooth muscle cells of the guinea-pig's taenia coli by the Goldman equation. *J. Physiol. (London)* **204,** 193-208.

Casteels, R. (1971). The distribution of chloride ions in the smooth muscle cells of the guinea-pig's taenia coli. *J. Physiol. (London)* **214,** 225-243.

Casteels, R., and Wuytack, F. (1975). Aerobic and anaerobic metabolism in smooth muscle cells of taenia coli in relation to active ion transport. *J. Physiol. (London)* **250,** 203-220.

Charnock, J. S., Cook, D. A., Almeida, A. F., and To, R. (1973). Activation energy and phospholipid requirements of membrane bound adenosine triphosphatases. *Arch. Biochem. Biophys.* **159,** 393-399.

Cox, R. H., Jones, A. W., and Fischer, G. M. (1974). Correlation of changes in carotid artery mechanics with connective tissue and electrolyte contents in puppies. *Am. J. Physiol.* **227,** 563-568.

Cox, R. H., Jones, A. W., and Swain, M. L. (1976). Mechanics and electrolyte composition of arterial smooth muscle in developing dogs. *Am. J. Physiol.* **231,** 77-83.

Davies, O. L. (1960). "Design and Analysis of Industrial Experiments." Hafner, New York.

Dunstone, J. R. (1962). Ion-exchange reactions between acid mucopolysaccharides and various cations. *Biochem. J.* **85,** 336-351.

Friedman, S. M., Mar, M., and Nakashima, M. (1974). Lithium substitution analysis of Na and K phases in a small artery. *Blood Vessels* **11,** 55-64.

Garay, R. P., Moura, A. M., Osborne-Pellegrin, M. J., Papadimitriou, A., and Worcel, M. (1979). Identification of different sodium compartments from smooth muscle cells, fibroblasts and endothelial cells, in arteries and tissue culture. *J. Physiol. (London)* **287,** 213-229.

Garrahan, P., Villamil, M. F., and Zadunaisky, J. A. (1965). Sodium exchange and distribution in the arterial wall. *Am. J. Physiol.* **209,** 955-960.

Goodford, P. J. (1968). Distribution and exchange of electrolytes in intestinal smooth muscle. *In* "Handbook of Physiology: Alimentary Canal" (C. F. Code, ed.), Vol. IV, Sect. 6, Chapter 86, pp. 1743-1766. Am. Physiol. Soc., Washington, D.C.

Heidlage, J. G., and Jones, A. W. (1981). The kinetics of active Na-K transport in the rabbit carotid artery. *J. Physiol. (London)* **317,** 243-262.

Hendrickx, H., and Casteels, R. (1974). Electrogenic sodium pump in arterial smooth muscle cells. *Pfluegers Arch.* **346,** 299-306.

Hermsmeyer, K. (1976). Electrogenesis of increased norepinephrine sensitivity of arterial vascular muscle in hypertension. *Circ. Res.* **38,** 362-367.

Hodgkin, A. L., and Horowicz, P. (1959). The influence of potassium and chloride ions on the membrane potential of single muscle fibres. *J. Physiol. (London)* **148,** 127-160.

Hoffman, P. G., and Tosteson, D. C. (1971). Active sodium and potassium transport in high potassium and low potassium sheep red cells. *J. Gen. Physiol.* **58,** 438-466.

Johansson, B., and Somlyo, A. P. (1980). Electrophysiology and excitation-contraction coupling. *In* "Handbook of Physiology: The Cardiovascular System, Vol. II, Vascular Smooth Muscle" (D. F. Bohr, A. P. Somlyo, and H. V. Sparks, eds.), pp. 301-323. Am. Physiol. Soc., Washington, D.C.

Jones, A. W. (1973). Altered ion transport in vascular smooth muscle from spontaneously hypertensive rats and influences of aldosterone, norepinephrine, and angiotensin. *Circ. Res.* **33,** 563-572.

Jones, A. W. (1974). Reactivity of ion fluxes in rat aorta during hypertension and circulatory control. *Fed. Proc.,* **33,** 133-137.

Jones, A. W. (1975). Analysis of bulk diffusion limited exchange of ions. *In* "Methods in Pharmacology. Smooth Muscle" (E. E. Daniel and D. M. Patton, eds.), Vol. III, Chapter 39, pp. 673-687. Plenum, New York.

Jones, A. W. (1976). Functional changes in vascular smooth muscle associated with experimental hypertension. *In* "Vascular Neuroeffector Mechanisms" (J. A. Bevan, G. Burnstock, B. Johansson, R. A. Maxwell, and O. A. Nedergaard, eds.), pp. 182-189. Karger, Basel.

Jones, A. W. (1980). Content and fluxes of electrolytes. *In* "Handbook of Physiology: The Cardiovascular System, Vol. II, Vascular Smooth Muscle." (D. F. Bohr, A. P. Somlyo, and H. V. Sparks, eds.), Sect. 2, pp. 253-299. Am. Physiol. Soc., Washington, D. C.

Jones, A. W. (1981a). Vascular smooth muscle and alterations during hypertension. *In* "Smooth Muscle: An Assessment of Current Knowledge" (E. Bülbring, A. F. Brading, A. W. Jones, and T. Tomita, eds.), pp. 397-429. Arnold, London.

Jones, A. W. (1981b). Kinetics of active sodium transport in aorta from control and deoxycorticosterone hypertensive rats. *Hypertension* (in press).

Jones, A. W. (1982). Ionic Dysfunction and Hypertension. *In* "Advances in Microcirculation, Vol. X, Ionic Regulation of the Microcirculation" (B. Altura, ed.). S. Karger, Basel.

Jones, A. W., and Hart, R. G. (1975). Altered ion transport in aortic smooth muscle during deoxycorticosterone acetate hypertension in the rat. *Circ. Res.* **37,** 333-341.

Jones, A. W., and Karreman, G. (1969a). Ion exchange properties of the canine carotid artery. *Biophys. J.* **9,** 884-909.

Jones, A. W., and Karreman, G. (1969b). Potassium accumulation and permeation in the canine carotid artery. *Biophys. J.* **9,** 910-924.

Jones, A. W., and Miller, L. A. (1978). Ion transport in tonic and phasic vascular smooth muscles and changes during hypertension. *Blood Vessels* **15,** 83-92.

Jones, A. W., and Swain, M. L. (1972). Chemical and kinetic analyses of sodium distribution in canine lingual artery. *Am. J. Physiol.* **223,** 1110-1118.

Jones, A. W., Somlyo, A. P., and Somlyo, A. V. (1973). Potassium accumulation in smooth muscle and associated ultrastructural changes. *J. Physiol. (London)* **232,** 247-273.

Jones, A. W., Sander, P. D., and Kampschmidt, D. L. (1977). The effect of norepinephrine on aortic ^{42}K turnover during deoxycorticosterone acetate hypertension and antihypertensive therapy in the rat. *Cir. Res.* **41,** 256-260.

Jonsson, O., Johansson, B., Wennergren, G., and Stage, L. (1975). Effects of temperature on osmotic responses and on transmembrane efflux of urea and sodium in vascular smooth muscle cells. *Experientia* **31,** 60-62.

Kamm, K. E., Zatzman, M. L., Jones, A. W., and South, F. E. (1979a). Maintenance of ion concentration gradients in the cold in aorta from rat and ground squirrel. *Am. J. Physiol.* **237,** C17–C22.

Kamm, K. E., Zatzman, M. L., Jones, A. W., and South, F. E. (1979b). Effects of temperature on ionic transport in aortas from rat and ground squirrel. *Am. J. Physiol.* **237,** C23–C30.

Krantz, J. C., Carr, C. J., and Knapp, M. J. (1951). Alkyl nitrites. XV. The effect of nitrites and nitrates on the oxygen uptake of arterial tissue. *J. Pharmacol. Exp. Ther.* **102,** 258–260.

Lehninger, A. L. (1971). "Bioenergetics," 2nd ed. Benjamin, Menlo Park, California.

MacGregor, E. A., and Bowness, J. M. (1971). Interaction of proteoglycans and chondroitin sulfates with calcium or phosphate ions. *Can. J. Biochem.* **49,** 417–425.

Manery, J. F. (1954). Water and electrolyte metabolism. *Physiol. Rev.* **34,** 334–417.

Mathews, M. B. (1975). "Connective Tissue: Macromolecular Structure and Evolution," Chapter 6. Springer-Verlag, Berlin and New York.

Merrillees, N. C. R. (1968). The nervous environment of individual smooth muscle cells of the guinea-pig vas deferens. *J. Cell Biol.* **37,** 794–817.

Miller, L. (1977). A study of ionic fluxes in tonic and phasic vascular smooth muscle of the rabbit and the effects of vasoactive agents. Ph.D. Thesis, University of Missouri, Columbia.

Nicholson, G. L. (1976). Transmembrane control of the receptors on normal and tumor cells. I. Cytoplasmic influence over cell surface components. *Biochim. Biophys. Acta* **457,** 57–108.

Overbeck, H. W., Pamnani, M. B., Akera, T., Brody, T. M., and Haddy, F. J. (1976). Depressed function of a ouabain-sensitive sodium–potassium pump in blood vessels from renal hypertensive dogs. *Circ. Res.* **38,** Suppl. 2 48–52.

Palatý, V., Gustafson, B. K., and Friedman, S. M. (1971). Maintenance of the ionic composition of the incubated artery. *Can. J. Physiol. Pharmacol.* **49,** 106–112.

Paul, R. J. (1980). Chemical energetics of vascular smooth muscle. *In* "Handbook of Physiology: The Cardiovascular System, Vol. II, Vascular Smooth Muscle" (D. F. Bohr, A. P. Somlyo, and H. V. Sparks, eds.), Sect. 2, pp. 201–235. Am. Physiol. Soc., Washington, D.C.

Raison, J. K., Lyons, J. M., and Thomson, W. W. (1971). The influence of membranes on the temperature-induced changes in the kinetics of some respiratory enzymes of mitochondria. *Arch. Biochem. Biophys.* **142,** 83–90.

Serafini-Fracassini, A., and Smith, J. W. (1974). "The Structure and Biochemistry of Cartilage." Churchill-Livingstone, Edinburgh and London.

Shuman, H., Somlyo, A. V., and Somlyo, A. P. (1976). Quantitative electron probe microanalysis of biological thin sections: Methods and validity. *Ultramicroscopy* **1,** 317–339.

Shuman, H., Somlyo, A. V., and Somlyo, A. P. (1977). Theoretical and practical limits of E_d x-ray analysis of biological thin sections. *Scanning Electron Microsc.* **1,** 663–672.

Snedecor, G. W., and Cochran, W. G. (1967). "Statistical Methods," 6 ed. Iowa State Univ. Press, Ames.

Somlyo, A. P., and Somlyo. A. V. (1968). Vascular smooth muscle. I. Normal structure, pathology, biochemistry and biophysics. *Pharmacol. Rev.* **20,** 197–272.

Somlyo, A. P., and Somlyo, A. V. (1970). Vascular smooth muscle. II. Pharmacology of normal and hypertensive vessels. *Pharmacol. Rev.* **22,** 249–353.

Somlyo, A. P., Somlyo, A. V., and Shuman, H. (1979). Electron probe analysis of vascular smooth muscle: Composition of mitochondria, nuclei and cytoplasm. *J. Cell Biol.* **81,** 316–335.

Somlyo, A. V., Vinall, P., and Somlyo, A. P. (1969). Excitation–contraction coupling and electrical events in two types of vascular smooth muscle. *Microvasc. Res.* **1,** 354–373.

Thomas, R. C. (1972). Electrogenic sodium pump in nerve and muscle cells. *Physiol. Rev.* **52,** 563–594.

Tobian, L. (1960). Interrelationship of electrolytes, juxtaglomerular cells and hypertension. *Physiol. Rev.* **40,** 280–312.

Van Breemen, C., Aaronson, P., Loutzenhiser, R., and Meisheri, K. (1980). Ca^{2+} movements in smooth muscle. *Chest* **78,** 157–173.

Villamil, M. F., and Matloff, J. (1975). Changes in vascular ionic content and distribution across aortic coarctation in the dog. *Am. J. Physiol.* **228,** 1087–1093.

Villamil, M. F., Rettori, V., Barajas, V., and Kleeman, C. R. (1968a). Extracellular space and the ionic distribution in the isolated arterial wall. *Am. J. Physiol.* **214,** 1104–1112.

Villamil, M. F., Rettori, V., Yeyati, N., and Kleeman, C. R. (1968b). Chloride exchange and distribution in the isolated arterial wall. *Am. J. Physiol.* **215,** 833–839.

Webb, R. C., and Bohr, D. F. (1979). Potassium relaxation of vascular smooth muscle from spontaneously hypertensive rats. *Blood Vessels* **16,** 71–79.

Wei, J. W., Janis, R. A., and Daniel, E. E. (1976). Calcium accumulation and enzymatic activities of subcellular fractions from aortas and ventricles of genetically hypertensive rats. *Circ. Res.* **39,** 133–140.

3

Membrane Electrical Activation of Arterial Smooth Muscle

David R. Harder

I. INTRODUCTION

In recent years there have been a number of good reviews discussing what is known about the electrical properties of vascular smooth muscle (Kumamoto, 1977; Horn, 1978; Johansson, 1978a; Bolton, 1979). This chapter will not attempt to review the literature again, but will address itself to some specific aspects regarding membrane regulation of the contraction of vascular smooth muscle and, more specifically, arterial smooth muscle.

VASCULAR SMOOTH MUSCLE: METABOLIC, IONIC, AND CONTRACTILE MECHANISMS

ISBN 0-12-195220-7

Because of the heterogeneous responses of smooth muscle from different vessels to a wide variety of stimuli, it is difficult to make generalizations about the membrane properties of arterial smooth muscles as a group. However, the arterial smooth muscle cells from most arteries studied have in common a relatively low resting membrane potential (E_m) compared to other muscles, have a sizable electrogenic pump potential contribution to the E_m, and depolarize to an incremental increase in extracellular K^+ (below 100 m*M*) to a lesser degree than if the Nernst potential for a K^+-selective membrane (E_K) = $[-60 \log ([K]_i/[K]_o)]$ were followed. These characteristics can account for some of the unique properties of arterial smooth muscle and vascular smooth muscle in general, as will be discussed in Section II.

Many of our present concepts regarding mammalian vascular smooth muscle electrophysiology come from comparisons of the portal or anterior mesenteric vein with various arteries of different animal species. Such studies have led to the classification of vascular smooth muscle in terms of vessels that exhibit spontaneous electrical spike activity (action potentials) and those that do not. In general, only the anterior mesenteric or portal vein falls into the former category, whereas most arteries, with a few noticeable exceptions, fall into the latter when studied *in vitro*. Even arterial preparations that exhibit spontaneous electrical spike activity *in situ*, such as small mesenteric arteries, are electrically quiescent when studied as isolated segments (von Loh and Bohr, 1973). However, isolated arterial preparations are capable of generating action potentials when the outward K^+ conductance is inhibited by a variety of agents (to be discussed). Thus, arterial smooth muscle possesses the ionic channels necessary for regenerative electrical activity; however, they are masked, at least in an artificial *in vitro* environment, by a large outward K^+ current.

Depolarization of arterial smooth muscle increases the influx of Ca^{2+}, resulting in tension development. However, the degree to which changes in E_m are related to changes in tension remains somewhat obscure. This is due, in part, to the ability of arterial segments to contract in the absence of a large or noticeable depolarization of the smooth muscle cells when stimulated by certain vasoactive agents. For example, norepinephrine, in very low or high concentrations, contracts certain arterial preparations in the absence of a change in E_m (Su *et al.*, 1964; Somlyo and Somlyo, 1968a; Casteels *et al.*, 1977b; Droogmans *et al.*, 1977). As will be discussed, the existence of a nonelectrical mechanism for activation does not necessarily demonstrate that the major activation process in arterial smooth muscle is nonelectrical.

II. MEMBRANE POTENTIAL (E_m) OF ARTERIAL SMOOTH MUSCLE AS A FUNCTION OF EXTERNAL K^+ ($[K]_o$)

A. E_m Versus Log $[K]_o$ Curves, and P_{Na}/P_K Ratio

The E_m in arterial smooth muscle ranges between −35 and −60 mV (see reviews by Kumamoto, 1977; Horn, 1978). As one increases $[K]_o$ beyond 10 m*M*, arterial smooth muscle depolarizes as the K^+ equilibrium potential (E_K) decreases. The average slope of the E_m versus log $[K]_o$ curve (measured between 10 and 100 m*M* $[K]_o$) ranges between 32 and 48 mV/decade (Kuriyama and Suzuki, 1978a,b; Hermsmeyer, 1976a; Harder and Sperelakis, 1978; Harder and Coulson, 1979). Thus, increasing $[K]_o$ does not depolarize arterial smooth muscle as if it behaved as a purely K^+-selective membrane as predicted from the Nernst equation for a 10-fold change in $[K]_o$ (Fig. 1), i.e., the slope does not approximate 60 mV/decade.

The relatively low slope of the E_m versus log $[K]_o$ curve at $[K]_o$ values greater than 10 m*M* in arterial smooth muscle can be related to at least two factors. First, both the calculated

$$\frac{P_{Na}}{P_K} = \frac{[K]_i - [K]_o \text{ (antilog } E_m/-60 \text{ mV)}}{[Na]_o \text{ (antilog } E_m/-60 \text{ mV)} - [Na]_i}$$

(Harder and Sperelakis, 1978) and measured (Casteels *et al.*, 1977a) ratios of Na^+ permeability (P_{Na}) to K^+ permeability (P_K) in arterial smooth muscle is about 0.2. This value is much greater than the value of about 0.01–0.05 found in skeletal muscle. The probable reason for this high P_{Na}/P_K ratio in arterial smooth muscle is related to a low P_K rather than a high P_{Na}. The reasons for this assumption include a high input resistance in arterial smooth muscle (Mekata, 1974; Harder and Sperelakis, 1978) and the fact that input resistance is essentially unaffected by inhibition of Na^+ influx by amiloride (Harder and Sperelakis, 1978). Not only do the high P_{Na}/P_K ratio and low P_K account for the relatively low E_m in arterial smooth muscle, but they also account for the relative flatness of the E_m versus log $[K]_o$ curve at $[K]_o$ values between 10 and 100 m*M*, in that they cause a deviation from E_K.

Second, as $[K]_o$ is elevated in vascular smooth muscle, K^+ conductance (g_K) increases (Hermsmeyer, 1976c). (The conductance of an ion through a membrane is the reciprocal of its resistance and is related to its ability to carry current. It is defined as the net current flow per unit

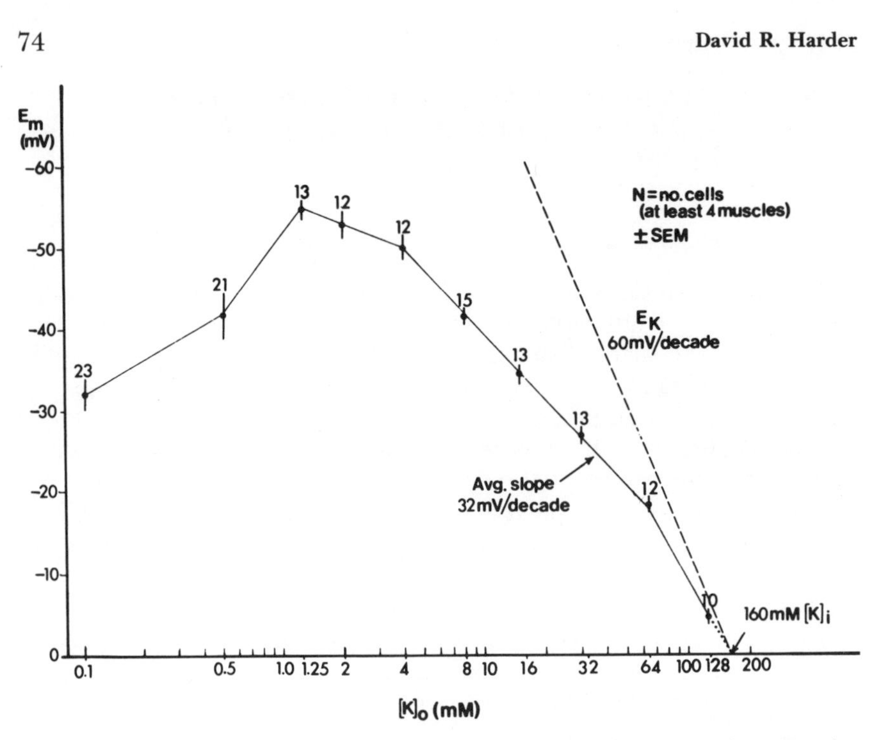

Fig. 1. Resting potential (E_m) as a function of external K^+ concentration ($[K]_o$) (log scale) for vascular smooth muscle of guinea pig superior mesenteric artery. The vertical bars represent the mean ± 1 SEM for the number of cells indicated in each case; the data for each point plotted were collected from four to six muscles. The curve extrapolated to zero potential gives an estimated internal K^+ concentration of 160 m*M*. The broken line gives the K^+ equilibrium potential (E_K), as calculated from the Nernst equation, and has a slope of 60 mV per 10-fold change in $[K]_o$. Reproduced by permission, Harder and Sperelakis (1978).

voltage. For example, if g_K increases, the amount of current carried by K^+ increases.) Such an increase in g_K acts to hyperpolarize the membrane at any given elevation in $[K]_o$. Thus the amount of depolarization at any given $[K]_o$ value is less because of the effect of K^+ on g_K. This factor tends to be more prominent in arterial smooth muscle cells which have a low g_K and P_K, since it is further from E_K (see Sperelakis, 1980, for a more complete discussion).

High g_K in Middle Cerebral Arteries of the Cat

Increasing $[K]_o$ depolarizes the middle cerebral artery with a slope of 58 mV/decade (between 10 and 100 m*M* $[K]_o$) versus 36 mV/decade in the mesenteric artery of the same animal (Fig. 2). The slope of 58 mV/decade approaches the Nernst potential for a K^+-selective membrane

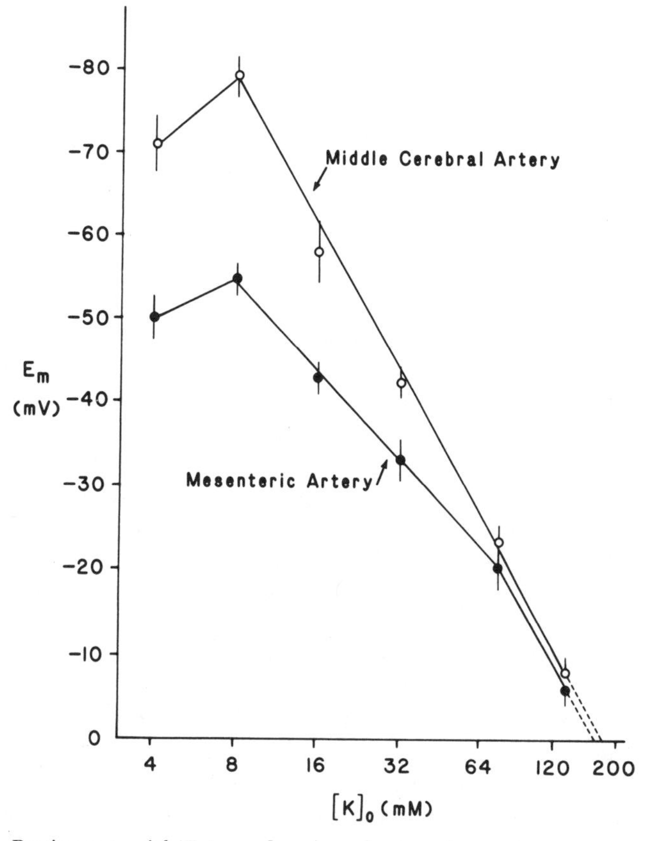

Fig. 2. Resting potential (E_m) as a function of external potassium concentration ($[K]_o$) (log scale) for vascular smooth muscle of cat middle cerebral and mesenteric arteries. The vertical bars represent the $E_m \pm$ SEM for at least seven different cells in at least four separate arteries. The average slope of the E_m versus log $[K]_o$ curve (between 10 and 100 mM $[K]_o$) in the middle cerebral artery. Extrapolation to zero potential (dashed lines) gives an estimated $[K]_i$ value of 180 mM for the middle cerebral artery and 170 mM for the mesenteric artery. Reproduced by permission, Harder (1980a).

and is probably due to an increased g_K compared to that in other arteries, since the input resistance of the membrane is also significantly lower (Harder, 1980a).

A higher g_K in the middle cerebral artery makes it more sensitive to small changes in $[K]_o$ than other arteries, i.e., a greater change occurs in E_m for a given change in $[K]_o$. Such data suggest that, in some cerebral arteries, a greater change in E_m may be required to affect a change in tension. This may alter the sensitivity of some cerebral arteries to vasoactive agents but may make them more sensitive to the K^+ concentration in their environment.

B. Stimulation of the Electrogenic Na^+–K^+ Pump by Changing $[K]_o$

Lowering $[K]_o$ to less than 2 m*M* depolarizes arterial smooth muscle cells. This depolarization is due to inhibition of an electrogenic Na^+–K^+ pump and perhaps a reduction in g_K (Harder and Sperelakis, 1978). Upon raising $[K]_o$ after an artery has been in low-K^+ solutions, significant hyperpolarization of the membrane is recorded, which is probably due to stimulation of this electrogenic Na^+–K^+ pump (Chen *et al.*, 1972; Bonaccorsi *et al.*, 1977; Haddy, 1975, 1978). Upon maximum stimulation of the electrogenic Na^+–K^+ pump, a hyperpolarization of up to 27 mV occurs (Bonaccorsi *et al.*, 1977). The reason low $[K]_o$ followed by high $[K]_o$ stimulation of the electrogenic pump is so effective is related to the fact that when the membrane is depolarized by inhibition of Na^+, K^+-ATPase in low-K^+ solutions, there is an increase in intracellular Na^+. In addition, Na^+,K^+-ATPase is not maximally stimulated until the $[K]_o$ reaches 6–10 m*M* (Schied and Fay, 1980).

This K^+-induced hyperpolarization is closely linked to relaxation in arteries (Chen *et al.*, 1972; Webb and Bohr, 1978a; Bonnacorsi *et al.*, 1977; Haddy, 1975, 1978) and has proved to be a useful tool in the study of disease states which may involve abnormalities of vascular smooth muscle, such as hypertension. Hermsmeyer (1976b) has shown that, in arteries from spontaneously hypertensive rats, a greater portion of the E_m is contributed by an electrogenic Na^+–K^+ pump potential as compared to the situation in normotensive controls. Such a hypothesis has gained support from the findings of Webb and Bohr (1979) showing a greater K-induced relaxation in spontaneously hypertensive rats. The reason for a greater electrogenic Na^+–K^+ pump in these animals appears to be a compensatory mechanism for a decreased E_K and possibly an increased intracellular Na^+ content (Hermsmeyer, 1976b; Jones, 1973).

The Na^+–K^+ pump appears to be suppressed in both cardiac and vascular smooth muscle of animals with experimental low-renin, presumably volume-expanded hypertension. The evidence for reduced Na^+–K^+ pump activity includes decreased K-induced vasodilation (Overbeck and Haddy, 1967; Overbeck, 1972), decreased ouabain-sensitive ^{86}Rb uptake by blood vessels (Overbeck *et al.*, 1976; Pamnani *et al.*, 1978), and decreased Na^+,K^+-ATPase activity in cardiac microsomes (Clough *et al.*, 1977). Since the pump is electrogenic, it has been suggested that decreased pump activity leads to depolarization, increased

Ca^{2+} permeability and influx, vasoconstriction, and increased sensitivity to vasoactive agents (see reviews by Haddy and Overbeck, 1976; Haddy *et al.*, 1978; Van Breeman *et al.*, 1979). On the other hand, Blaustein (1977) has suggested that decreased pump activity leads to vasoconstriction via the Na^+–Ca^{2+} exchange mechanism (rather than via electrogenic depolarization).

Thus, the way in which the arterial smooth muscle cell responds to changes in $[K]_o$ can account for some of the unique characteristics of the muscle and has proved to be a useful tool in the study of hypertension.

III. K^+ CONDUCTANCE (g_K) AND ELECTRICAL ACTIVITY

A. Agents That Alter g_K

When isolated arteries are impaled with intracellular microelectrodes, they are usually quiescent. Spontaneous electrical spike (action potential) generation usually cannot be recorded, even with the application of strong intra- or extracellular current pulses (Harder and Sperelakis, 1978; Harder *et al.*, 1979; Casteels *et al.*, 1977a; Droogmans *et al.*, 1977; Mekata, 1971, 1974; Holman and Surprenant, 1979; Hermsmeyer, 1971; Mekata and Niu, 1972; Kuriyama and Suzuki, 1978a,b; Ito *et al.*, 1978a). However, when agents that decrease g_K are added to such preparations, action potentials can be recorded either upon electrical stimulation or spontaneously, depending upon the dose.

Barium (Ba^{2+}) at doses of 10^{-4} M and above depolarizes and induces spontaneous electrical activity in arterial smooth muscle (Harder and Sperelakis 1978, 1979). Ba^{2+} is a potent inhibitor of outward K^+ current and markedly increases membrane resistance, as indicated by an increased slope of the voltage versus current curve (Fig. 3). In the guinea pig mesenteric artery, Ba^{2+} rapidly induces depolarization and spontaneous action potentials in this previously quiescent preparation (Fig. 4). The magnitude of the Ba^{2+} depolarization is time-dependent, continually depolarizing arterial smooth muscle, an action perhaps related to its entry into the cell (Fig. 4). Induction of electrical activity with Ba^{2+} may not be solely related to its action on decreasing g_K. Uvelius *et al.* (1974) have shown that Ba^{2+} can substitute for Ca^{2+} during the action potential in the portal vein.

Procaine and other local anesthetics induce or enhance electrical activity in a wide variety of smooth muscle preparations (see review by Bolton, 1979). This effect of procaine appears to be due to its action in

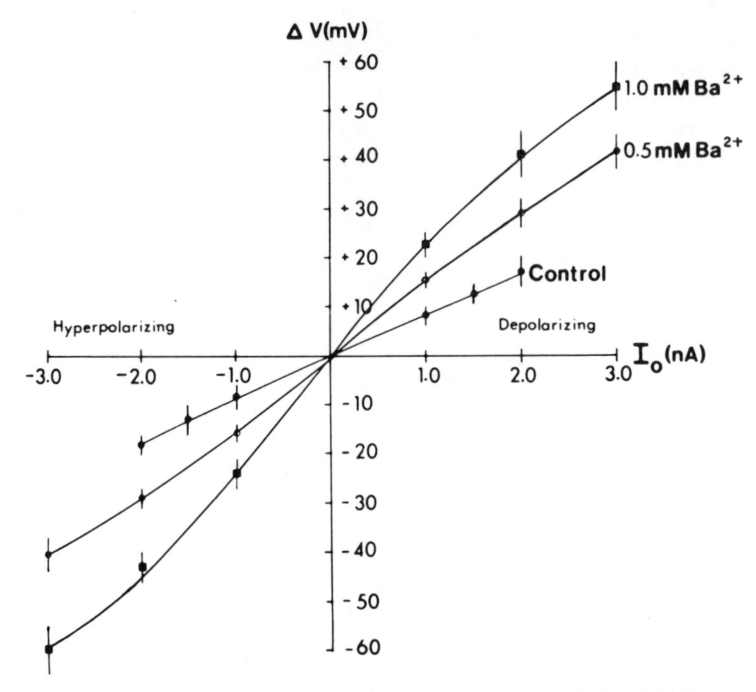

Fig. 3. Steady-state voltage versus applied current (I_0) relationship for vascular smooth muscle of guinea pig superior mesenteric artery bathed in normal Ringer's solution and in a solution containing Ba^{2+} (0.5 and 1.0 m*M*). Each point represents the change in E_m (ΔV) in response to rectangular hyperpolarizing or depolarizing current pulses applied through the recording microelectrode. The vertical bars represent ± SEM for at least 10 cells in at least four different preparations. The slope through the origin (zero applied current) gives the input resistance. Note that Ba^{2+} increases the slope and therefore the input resistance. Reproduced by permission, Harder and Sperelakis (1978).

reducing g_K (Jacobs and Keatinge, 1974; Kuriyama and Suzuki, 1978a; Casteels *et al.*, 1977a).

Tetraethylammonium ion (TEA) is probably the best known of the agents that reduce g_K and induce rapid regenerative electrical activity in previously quiescent preparations (see review by Kumamoto, 1977). This ion induces action potential generation in guinea pig mesenteric arteries (Harder and Sperelakis, 1978), canine coronary arteries (Harder *et al.*, 1979), rabbit common carotid artery (Mekata, 1971), rabbit ear artery (Droogmans *et al.*, 1977), and rabbit saphenous artery (Holman and Surprenant, 1979). Characterization of the TEA action potential will be more fully discussed in Section III,B.

Thus, it appears that arterial smooth muscle *in vitro* is capable of generating action potentials if g_K is reduced. Furthermore, action poten-

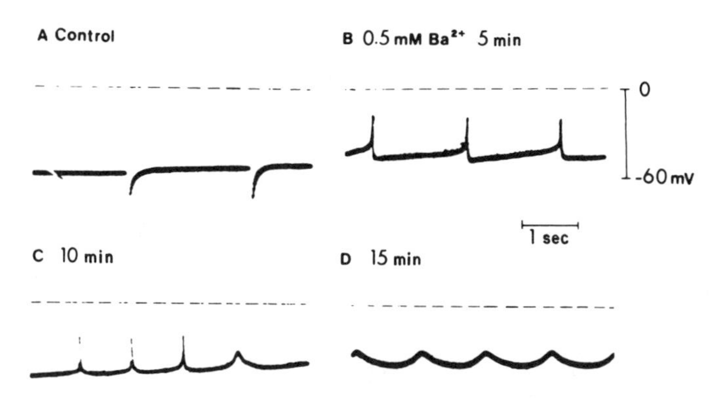

Fig. 4. Induction of excitability by Ba^{2+} in guinea pig superior mesenteric artery muscle. (A) Control record showing a large resting potential of about − 56 mV and a lack of spontaneous action potentials or responses to external electrical stimulation (two shock artifacts shown). (B) Record taken from same cell impaled in (A) 5 min after the addition of 0.5 m*M* Ba^{2+}, illustrating the production of spontaneous action potentials and partial depolarization. (C and D) Records taken from another cell 10 min (C) and 15 min (D) after the addition of Ba^{2+}, illustrating the further depolarization and shift from action potentials to small oscillations. The voltage and time calibration in (B) apply throughout. Reproduced by permission, Harder and Sperelakis (1978).

tials can be induced upon electrical stimulation in the presence of a dose of TEA that has no effect on E_m, suggesting that it inhibits voltage-sensitive as well as resting g_K (Harder and Sperelakis, 1979). Such findings suggest that the outward K^+ current of arterial smooth muscle is sufficiently rapid so as to mask or overlap the relatively slow inward current in arterial smooth muscle, the net result of which is the absence of rapid regenerative electrical activity. As will be discussed (Section III,B), the current carried during the TEA action potential is primarily Ca^{2+}. With this in mind, the action of TEA on reducing the outward current, thereby allowing regenerative electrical activity, is illustrated in Fig. 5.

Ca^{2+}-Induced Increase in g_K

When intracellular Ca^{2+} is elevated in isolated amphibian smooth muscle, g_K is increased; g_K is decreased when free intracellular Ca^{2+} is reduced (Caffrey and Anderson, 1979; Singer and Walsch, 1980). Such findings suggest that g_K is regulated in part by intracellular Ca^{2+}, similar to the phenomena described in nerve (Meech, 1974) and cardiac tissue (Isenberg, 1977).

In canine coronary arteries, histamine increases inward Ca^{2+} current and decreases the input resistance of the membrane with concomitant

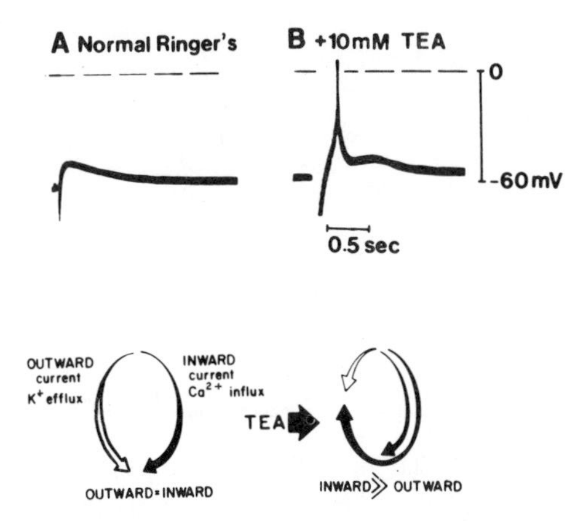

Fig. 5. Induction of excitability by TEA in the smooth muscle of a small coronary artery of the dog. Top: (A) Control in normal Ringer's solution showing inexcitability. (B) Record from the same cell as in (A) 10 min after the addition of 10 m*M* TEA, illustrating a large overshooting action potential in response to electric stimulation. Bottom: A hypothetical mechanism for TEA induction of regenerative spikes. Normally (without TEA), the outward K^+ current that flows upon depolarizing stimuli has a magnitude comparable to that of the inward Ca^{2+} current (open and closed arrows are equal); thus the vascular smooth muscle cells are inexcitable. The addition of TEA might alter the balance of inward and outward current either by reducing the outward K^+ current (smaller open arrow) and/or increasing the inward Ca^{2+} current (larger solid arrow). In either case, sufficient net inward current would flow, allowing generation of action potentials. Reproduced by permission of the American Heart Association, Inc., Harder *et al.* (1979).

hyperpolarization, suggesting an increase in g_K (Harder, 1980). When Ca^{2+} influx is blocked by Mn^{2+}, the histamine-induced decrease in input resistance is blocked (Fig. 6). These findings suggest that an increase in intracellular Ca^{2+} is capable of augmenting g_K in arterial smooth muscle, similar to isolated amphibian smooth muscle. If one assumes that, upon raising extracellular Ca^{2+}, intracellular Ca^{2+} will also increase because of an increase in driving force, then an increase in g_K may be partly responsible for the well-known "stabilizing" or hyperpolarizing action upon raising extracellular Ca^{2+}. Similarly, an increase in input resistance and depolarization can be recorded upon decreasing extracellular Ca^{2+} below normal values in rabbit pulmonary artery (Casteels *et al.*, 1977a). However, as pointed out by Casteels *et al.* (1977a), factors in addition to or instead of a decrease in P_K may be responsible for the depolarization in low-Ca^{2+} solutions. These factors may include a modulation of the Na^+–K^+ pump as a function of $[Ca]_o$ as proposed by Webb and Bohr (1978b).

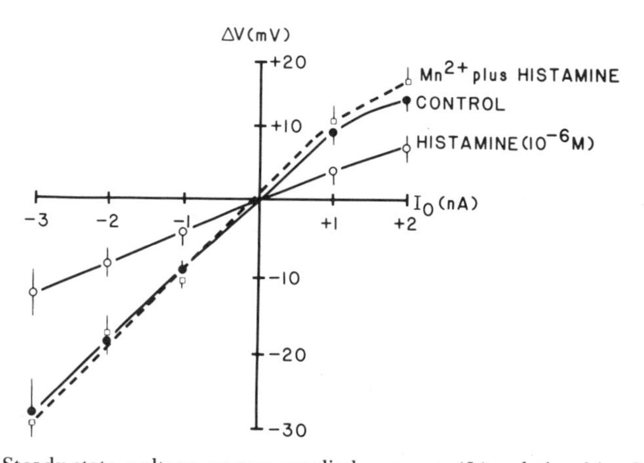

Fig. 6. Steady-state voltage versus applied current (I_0) relationship for vascular smooth muscle of canine coronary artery bathed in normal Krebs' solution (solid circles) and in a solution containing histamine (10^{-6} *M*) (open circles) or manganese ion (1 m*M*) plus histamine (10^{-6} *M*) (squares). Each point represents the change in E_m (ΔV) in response to rectangular hyperpolarizing or depolarizing current pulses applied through the microelectrode. The vertical bars represent the mean ± SEM for at least eight cells in five different preparations. The slope through the origin (zero applied current) gives the input resistance. Note that histamine decreased the slope and therefore the input resistance. In the presence of manganese ion, histamine had no significant effect on the slope. Reproduced by permission of the American Heart Association Inc., Harder (1980b).

B. Characterization of the Tetraethylammonium Ion-Induced Action Potential

The action potentials generated upon electrical stimulation in the presence of TEA increase in amplitude and maximal rate of rise as a function of the extracellular Ca^{2+} concentration (Fig. 7). The curve relating action potential amplitude versus log extracellular Ca^{2+} concentration has a slope of 29–30 mV/decade (between 0.5 and 5 m*M* $[Ca]_o$, Fig. 8). This is in nearly perfect agreement with the Nernst potential for a membrane selective for a divalent cation such as Ca^{2+}. These findings, together with observations that variations in extracellular Na^+ had no effect on either action potential amplitude or maximal rate of rise (Harder and Sperelakis, 1979), demonstrate that the inward current carried during this action potential is carried solely by Ca^{2+}. Identical findings were obtained in canine coronary artery (Harder *et al.*, 1979).

The fact that the inward current of the TEA-induced action potential is carried by Ca^{2+} makes it a valuable tool for the study of inward Ca^{2+} current in vascular smooth muscle. Even though this action potential is

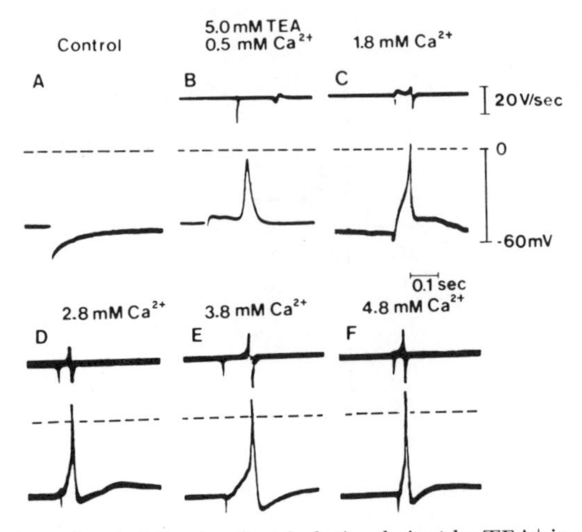

Fig. 7. Induction of excitability (to electrical stimulation) by TEA^+ in normally inexcitable vascular smooth muscle from guinea pig superior mesenteric artery. All records taken from one impalement. (A) Control record showing a large resting potential of −58 mV and a lack of spontaneous action potentials or responses to intense external electrical stimulation (one shock artifact depicted). (B–F) Production of action potentials (in response to electrical stimulation) after the addition of 5 m*M* TEA, illustrating an increase in amplitude and the maximal rate of rise ($+\dot{V}_{max}$) of action potentials as Ca^{2+} is increased from 0.5 m*M* (B) to 1.8 m*M* (C), 2.8 m*M* (D), 3.8 m*M* (E), and 4.8 m*M* (F). The upper trace in (B–F) gives dV/dt, the maximal deflection of which is proportional to $+\dot{V}_{max}$. The voltage, time, and $+\dot{V}_{max}$ calibrations in (C) apply throughout. The horizontal broken line gives the zero potential level. Reproduced by permission, Harder and Sperelakis (1979).

only an experimental model, it demonstrates that arterial smooth muscle cell membranes have the ionic channels required for action potential production and can be blocked by known Ca^{2+} antagonists (Harder and Sperelakis, 1979; Harder *et al.*, 1979).

The channels which carry the inward current during the TEA-induced action potential have some of the characteristics of channels which carry the slow inward current in embryonic and adult cardiac muscle (Sperelakis and Shigenobu, 1972; New and Trautwein, 1972). These action potentials have a slow rate of rise (less than 10 V/sec) and exhibit a voltage inactivation curve very similar to that of cardiac slow channels (Fig. 9), demonstrating complete voltage inactivation at − 22 mV.

Action potentials induced by TEA have been used to study the effect of a variety of vasoactive agents on Ca^{2+} inward current in a number of different arteries. In small (< 500 μm OD) canine coronary arteries, the Ca^{2+}-dependent TEA-induced action potential is blocked by adenosine but unaffected by nitroglycerine, whereas in large (> 1.0 mm

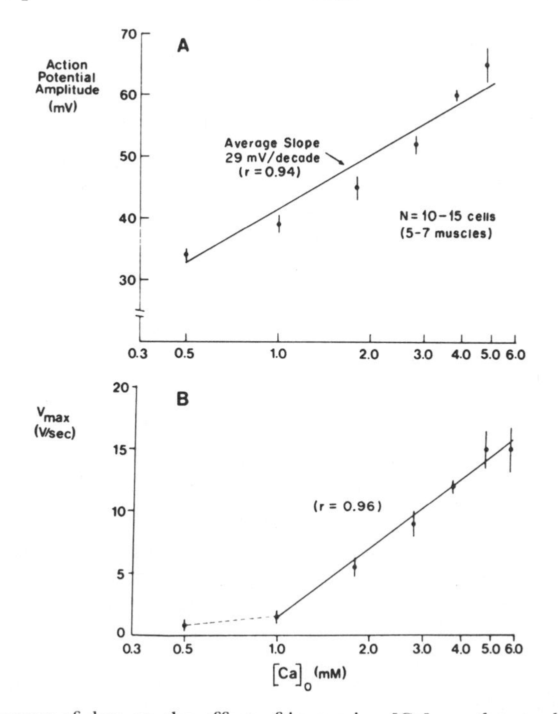

Fig. 8. Summary of data on the effect of increasing $[Ca]_o$ on the amplitude of TEA-induced action potentials (A) and on $+\dot{V}_{max}$ (B) in isolated vascular smooth muscle from guinea pig mesenteric arteries. Each point plotted is mean +SEM for 10–15 impalements in 5–7 different arteries. (A) Average slope of the action potential amplitude versus log $[Ca]_o$ curve is 29 mV/decade between 0.5 and 5.0 m*M* $[Ca]_o$ (correlation coefficient of 0.94). (B) Linear relationship between $+\dot{V}_{max}$ and $[Ca]_o$ between 1.0 and 5.0 m*M* $[Ca]_o$ (correlation coefficient of 0.96). At 0.5 m*M* $[Ca]_o$, the point fell off the calculated straight line. Curves in (A) and (B) were calculated using linear regression analysis. Note that abscissas are on log scales. Reproduced by permission, Harder and Sperelakis (1979).

OD) coronary arteries, adenosine does not affect the action potential but is blocked by nitroglycerine (Harder *et al.*, 1979). These differential actions of adenosine and nitroglycerine were demonstrated by Schnaar and Sparks (1972) on the contractions of large and small coronary arteries. Furthermore, Schnaar and Sparks (1972) suggested that adenosine and nitroglycerine relaxed coronary arteries by inhibiting Ca^{2+} influx. Such findings demonstrate the usefulness of the TEA-induced action potential for the study of Ca^{2+} inward current which may contribute to or stimulate the release of activator Ca^{2+} for contraction. However, Dutta *et al.* (1980) could not show that adenosine decreased Ca^{2+} influx in coronary arteries. The TEA action potential may also be affected by agents that do not modify Ca^{2+} influx and that may work

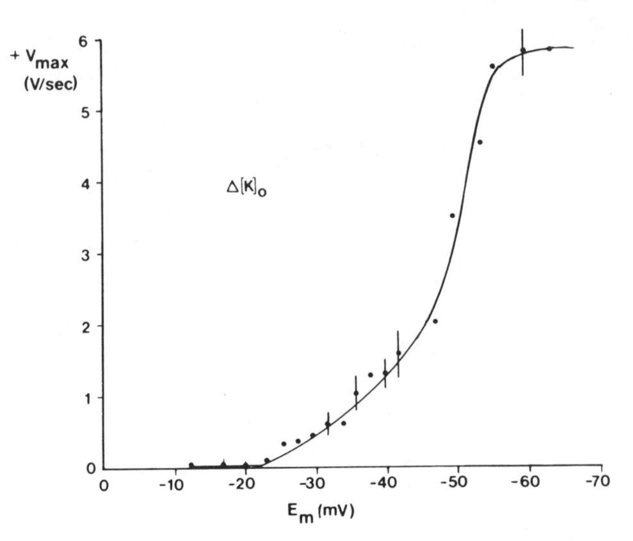

Fig. 9. Voltage inactivation curve of TEA-induced action potentials in vascular smooth muscle from guinea pig mesenteric arteries. Each point without vertical bars represents a mean of two cells; points with vertical bars represent the mean ± SEM of three or more cells; data were collected from five arteries. $+\dot{V}_{max}$ was measured as a function of resting membrane potential (E_m), which was varied by increasing $[K]_o$. Total inactivation of the action potential occurred at about $-$ 22 mV, and 50% inactivation occurred at about $-$ 47 mV. The dose of TEA was 5 mM. Action potentials were induced by electrical stimulation. Reproduced by permission, Harder and Sperelakis (1979).

through another mechanism, such as increasing g_K. Also, in coronary arteries, Propafenone, a relatively new antiarrhythmic agent that inhibits inward Ca^{2+} current in cardiac muscle and dilates coronary arteries, blocks the TEA-induced action potential in a dose-dependent fashion (Fig. 10). This action of Propafenone demonstrates that the TEA-induced action potential provides a useful tool for determining the actions of new therapeutic agents on Ca^{2+} inward current in arterial smooth muscle. For a more complete discussion of the action of vasoactive agents on the TEA-induced action potential in coronary arteries, see the review by Berne *et al.* (1981).

In cerebral arteries, relatively low doses of TEA that do not affect resting membrane properties allow alpha-receptor stimulation by norepinephrine to induce electrical spike activity, where norepinephrine only caused graded E_m changes before TEA (Fig. 11). Such data again point out the importance of the kinetics of g_K turn-on to membrane excitability in arterial smooth muscle. It is not clear whether or not arteries generate electrical spike activity *in vivo*. This section has pointed

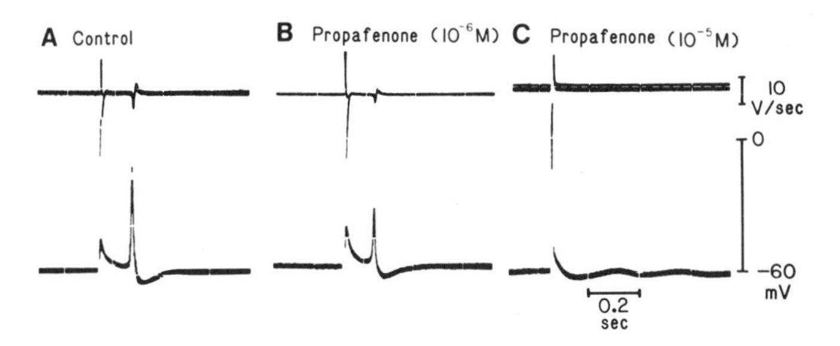

Fig. 10. Effect of Propafenone on the amplitude and maximal rate of rise ($+\dot{V}_{max}$) of the TEA-induced Ca^{2+}-dependent action potential in vascular smooth muscle of canine coronary arteries. (A) Control action potential induced by extracellular stimulation in the presence of 10 m*M* TEA. (B) Record from same cell showing a marked reduction in both the amplitude and $+\dot{V}_{max}$ of the TEA-induced action potential by 10^{-6} *M* Propafenone. (C) Complete inhibition of the action potential upon raising the concentration of Propafenone to 10^{-5} *M*. The voltage, time, and $+\dot{V}_{max}$ calibrations in (C) apply throughout. Reproduced by permission, Harder and Belardinelli (1980).

out the fact that arterial smooth muscle, even from large elastic arteries, is capable of generating action potentials under certain experimental conditions. When small mesenteric arteries are impaled with microelectrodes *in vivo*, spontaneous electrical activity is recorded (Steedman, 1966), but when studied *in vitro* the same arteries are quiescent (von Loh and Bohr, 1973). It is possible that, in the intact animal, humoral and/or nervous influences regulate excitability by modulating g_K.

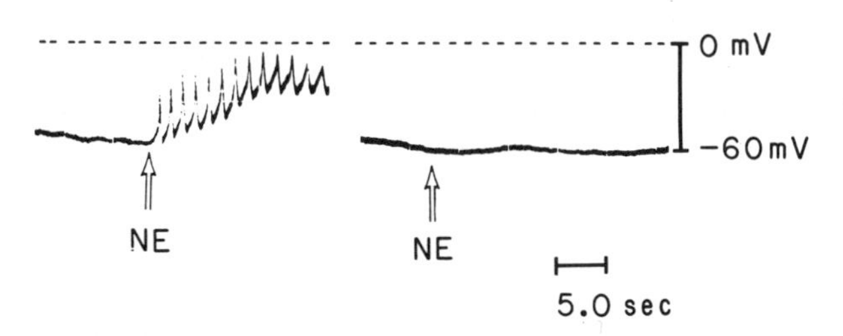

Fig. 11. Induction of excitability to application of 1μ*M* norepinephrine (NE) in muscle of basilar artery pretreated with 1 m*M* TEA, and abolition of this effect by phentolamine. (A) Electrical spike activity and depolarization induced by NE in an arterial segment pretreated with 1 m*M* TEA. (B) Addition of phentolomine 5 min before application of NE in a basilar artery pretreated with TEA. Voltage and time calibrations in (B) also apply to (A).

IV. IMPORTANCE OF E_m IN CONTROLLING THE ACTIVE STATE OF ARTERIAL SMOOTH MUSCLE

A. Graded Potential Changes

The low slope of the curve relating E_m to changes in $[K]_o$ may indicate that the development of tension in arterial smooth muscle is closely related to changes in E_m, in that even small changes in E_m are capable of inducing relatively large changes in tension. Siegel *et al.* (1976) have demonstrated a very close relationship between E_m and tension development in K^+-depolarized canine carotid arteries (between 0 and 10 m*M* $[K]_o$), with a depolarization of 6 mV causing a significant increase in tension. Ito *et al.* (1979) have also found a strong correlation between E_m and tension in K^+-depolarized rabbit pulmonary artery. Similarly, when the arterial smooth muscle membrane is depolarized by the application of outward current pulses, the developed tension is very closely related to changes in E_m, with only a 4-mV change in E_m resulting in significant tension development (Casteels *et al.*, 1977b). Thus, the level of the E_m in arterial smooth muscle is an important modulator of tension development. The amplitude of norepinephrine-induced contractions is significantly altered by changes in E_m (Casteels *et al.*, 1977b; Haeusler, 1978; Haeusler and Thorens, 1975). Depolarization of arterial smooth muscle by even several millivolts markedly increases tension development in response to norepinephrine (Casteels *et al.*, 1977b; Haeusler, 1978).

Hyperpolarization of arterial smooth muscle results in a decrease in tension and reduces the contractile response to vasoactive agents. In rabbit pulmonary artery sodium nitroprusside hyperpolarizes smooth muscle membranes, reduces baseline tension, and decreases the amount of tension development by K^+ and norepinephrine (Ito *et al.*, 1978b). Similarly, hyperpolarization of canine carotid arteries substantially inhibits the contractile response to norepinephrine (Haeusler, 1978). In cat pial arteries, hyperpolarization by norepinephrine induces relaxation in the absence of previously induced active tone (Harder, *et al.*, 1981). Such a finding is significant in that it is often thought that a certain level of previously existing tone is needed before an agent can relax vascular smooth muscle (Hester and Carrier, 1976) and shows that the E_m of arterial muscle cells is an important factor in regulating resting vessel tone.

Thus, it is well established that the level of the E_m in arterial smooth muscle is an important factor in regulating the contractile state. However, the physiological role of the E_m and its relationship to contraction mediated by certain neurotransmitters, such as norepinephrine, is not clearly established. In many of the isolated arterial preparations studied, contraction in response to low doses (10^{-9}–10^{-8} M) of norepinephrine is not preceded by membrane depolarization (Somlyo and Somlyo, 1968a; Casteels *et al.*, 1977a,b; Kuriyama and Suzuki, 1978b; Mekata and Niu, 1972; Droogmans *et al.*, 1977). Such contraction, not associated with depolarization of the smooth muscle cell, is commonly referred to as pharmacomechanical coupling (Somlyo and Somlyo, 1968a,b). The questions that arise when one attempts to delineate whether or not a particular vasoactive agent operates via electromechanical or pharmacomechanical coupling involve (1) which of the two mechanisms occurs *in vivo*, and (2) what mechanisms are involved in pharmacomechanical coupling.

It is generally believed that norepinephrine acts through two mechanisms when it induces tension in arterial smooth muscle: The first occurs at low doses (10^{-9}–10^{-7} M) at which it induces tension without changes in E_m, and the second occurs at higher doses (10^{-7}–10^{-5} M) at which the membrane depolarizes as tension increases (Casteels *et al.*, 1977b; Droogmans *et al.*, 1977; Mekata and Niu, 1972; Su *et al.*, 1964). Norepinephrine can also contract arteries that are already maximally depolarized (Haeusler, 1978; Somlyo and Somlyo, 1968b). The finding that low doses ($<10^{-8}$ M) of norepinephrine contract isolated arteries without depolarizing them has led to the rather widely believed assumption that this is the mechanism which occurs *in vivo* (Bohr, 1973), since these are the concentrations believed to be present in the environment of arterial smooth muscle cells in the intact animal. Such an assumption is perhaps premature in light of the fact that the dose dependency of norepinephrine differs among different arteries and different animals, and our knowledge of arterial smooth muscle electrophysiology is still rather limited in comparison to our knowledge of cardiac and skeletal muscle. For example, in rat pulmonary artery significant depolarization and contraction by norepinephrine occurs at 10^{-8} M, and in spontaneously hypertensive rats significant depolarization is observed at 10^{-9} M (Kuriyama and Suzuki, 1978a). In rabbit pulmonary artery, norepinephrine does not begin to depolarize until the concentration in the bathing solution reaches 10^{-7} M (Casteels *et al.*, 1977b). In cat basilar artery there is nearly perfect agreement between the dose–response curve for norepinephrine-induced contractions and depolarization at low doses of norepinephrine, with significant depolarization and accompanying contractions occurring at about 3×10^{-8} M (Harder *et al.*, 1981).

In rabbit pulmonary and ear arteries, the contractions recorded at low concentrations of norepinephrine, at which there is no change in E_m, are dependent upon an influx of Ca^{2+} from extracellular stores (Droogmans *et al.*, 1977; Casteels *et al.*, 1977b). Also, the permeability of Na^+, K^+, and Cl^- increases at doses of norepinephrine that do not depolarize the membrane (Casteels *et al.*, 1977a; Droogmans *et al.*, 1977). It has been postulated by Droogmans *et al.* (1977) that the reason the E_m does not change in certain arterial preparations, even through ion permeability increases upon exposure to norepinephine, is that there is a simultaneous increase in permeability of similar magnitude for all ionic species. Thus, even though norepinephrine does not induce depolarization at low concentrations, it does increase ionic conductance. Furthermore, before one can draw conclusions as to whether or not a certain agent contracts an arterial preparation by electrical or nonelectrical means, one must realize the difficulty in impaling arterial smooth muscle cells. Arterial muscle cells are small in size (3–5 μm OD) and are imbedded in a tough connective tissue matrix, thereby making the probability of poor cell impalement high. Poor impalements yield an artificially low value of E_m which may not allow one to see small changes in potential in response to a given vasoactive agent.

In rabbit pulmonary artery, the reduction of outward K^+ current by TEA reduces the minimal concentration of norepinephrine required to induce a significant depolarization from 5×10^{-8} to 5×10^{-9} M (Haeusler, 1978). Such findings suggest that at least one explanation as to why low concentrations of norepinephrine do not depolarize some arterial preparations is that there is an offsetting, hyperpolarizing outward current. It is possible that the induction of an outward K^+ current by norepinephrine is mediated by Ca^{2+} influx, as discussed in Section III,A,1. In support of this hypothesis is the observation that in the guinea pig coronary artery acetylcholine hyperpolarizes the membrane while still inducing an increase in tension (Kitamura and Kuriyama, 1979). This acetylcholine-induced hyperpolarization is due to an increase in g_K related to both an increased Ca^{2+} influx and a possible release of intracellularly bound Ca^{2+} (Kitamura and Kuriyama, 1979). Such an action of acetylcholine is not universal, however, since in porcine coronary artery acetylcholine increases tension without a change in E_m (Ito *et al.*, 1979).

Thus, it appears that the level of the E_m in arterial smooth muscle is an important determinant of tension development. Whether or not a particular artery exhibits electromechanical coupling does not appear to depend so much upon the particular vasoactive agent studied, but rather upon the particular artery and/or species from which it came. Such divergent findings make it difficult to say whether any given agent exerts its effects through electrical or nonelectrical coupling without qualification. Furthermore, excitatory junction potentials (EJP) can be recorded in response to perivascular stimulation of adrenergic nerves in blood vessels that did not depolarize to low doses of norepinephrine (Holman and Surprenant, 1979), demonstrating that low doses of neurotransmitter can indeed cause active membrane responses, but that vary markedly from the responses of exogenously applied neurotransmitter. Such findings suggest that exogenously applied transmitter may not reach receptors on the vascular smooth muscle cells due to diffusion barriers or neuronal uptake of norepinephrine, or they may suggest the existence of two receptor types: one that is responsive to neurally released transmitter and one which responds to that which is exogenously applied (Hirst and Neild, 1980).

B. Regenerative Electrical Events (Action Potentials)

It is relatively well accepted that in portal vein the frequency of electrical spike discharge controls, to a large extent, the amount of tension produced (Holman *et al.*, 1968; Johansson and Ljung, 1976; Johansson *et al.*, 1979; von Loh, 1971; Cuthbert, 1966; Funaki, 1966; Golenhofen and Heinstein, 1975; Golenhofen *et al.*, Hermsmeyer, 1973, 1976a; Somlyo and Somlyo, 1968a,b). There are only a few examples in the literature of arterial smooth muscle exhibiting electrical spike activity when studied *in vitro*. Somlyo *et al.* (1969) observed spontaneous electrical activity in a very small percent of the cases in rabbit main pulmonary artery. McLean and Sperelakis (1977) recorded spontaneous action potential generation in primary cultures of chick thoracic aorta.

In the cat middle cerebral artery, spontaneous electrical spike activity can be recorded in the presence of excess $[K]_o$ or serotonin (Fig. 12). The production of spontaneous electrical activity in the middle cerebral artery is one of the very few demonstrations of spontaneous spike generation in mammalian arteries *in vitro*. As mentioned previously, electrical spike activity can be recorded in mammalian arteries *in vivo*, but only rarely *in vitro*.

The function of such action potential generation in arterial smooth muscle is little understood. It is well documented that sustained depolarization of arterial smooth muscle results in the Ca^{2+} influx responsible for part of its contraction (Waugh, 1962; Briggs, 1962; Van Breemen, 1977; Blaustein, 1977; Johansson, 1978b; Van Breemen *et al.*, 1979). The contractions usually associated with such Ca^{2+} influx are slow, requiring several seconds.

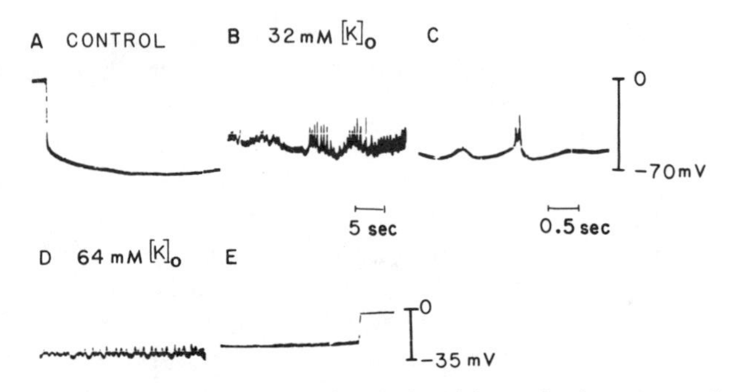

Fig. 12. Induction of spontaneous electrical activity and subsequent voltage inactivation in the smooth muscle cells of cat middle cerebral artery. (A) Control record showing large resting potential of −70 mV and lack of spontaneous action potentials. (B) Production of spontaneous electrical spike activity upon depolarizing the middle cerebral artery to about −48 mV with 32 m*M* $[K]_0$. (C) Record showing spontaneous electrical spike activity upon depolarization with 32 m*M* $[K]_o$ at an expanded oscilliscope sweep speed. (D) Record showing further depolarization with subsequent reduction in the amplitude of electrical spike activity taken 5 min after raising the K^+ concentration in the bath to 64 m*M*. (E) Record showing total voltage inactivation at about −25 mV taken 8 min after elevating $[K]_o$ to 64 m*M*. The membrane gradually depolarized upon elevating K^+ in the bathing solution, reaching a steady state within 8–10 min, thus accounting for the different resting potentials shown in records (D) and (E). Voltage calibration in (C) applies to (A) and (B), and in (E) applies to (D). Time calibrations in (B) apply to (A), (D), and (E). Records (A), (B), (D), and (E) are from the same cell (same impalement). Reproduced by permission, Harder (1980a).

The presence of action potentials brings up the exciting possibility that rapid contractions (on the order of a few hundred milliseconds) may be possible for arterial smooth muscle cells, similar to cardiac muscle. Such a concept is supported by the findings of Mulvany and Halpern (1977), showing that tension development in small arteries can occur in less than 1 sec, with a time to half-peak tension of 210 msec. The best evidence to date showing that arterial smooth muscle can contract with a time course closely following a simple action potential by Hermsmeyer (1979). He demonstrated that in primary cultures of rat aortic smooth muscle cells, initiation of the action potential preceded contraction by about 50 msec and that both the twitchlike contraction and the action potential had a total time course of about 200 msec (Fig. 13). McLean and Sperelakis (1977) have also observed rapid contractions from cultured vascular smooth muscle associated with action potentials.

The recording of an electrically mediated twitchlike contraction in arterial smooth muscle is a significant finding in vascular smooth muscle physiology, which may lead to a reevaluation of our current concepts

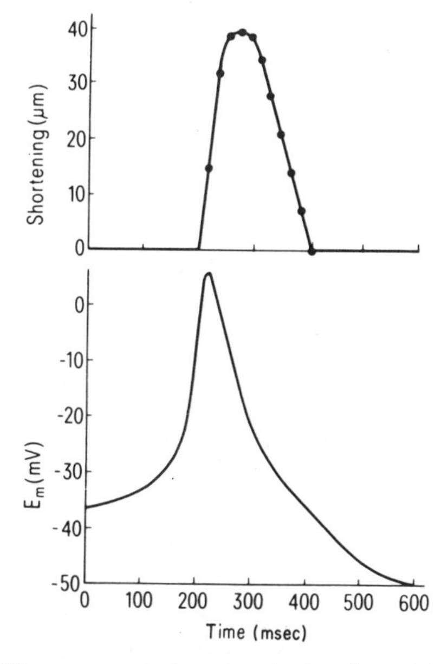

Fig. 13. A typical 200-msec contraction record taken from a (projected) single-frame film analysis and the spontaneous membrane excitation underlying it. The total length of this synchronously contracting small cell group was 440 μm. Contractile time course and spike duration at −30 mV were nearly equal. Contraction always began during the steep rising phase of the spike. Contractions of the S type were about 4 sec in duration, caused by spontaneous slow depolarization of 10–20 mV, and were superimposable on the shortening curve. Reproduced by permission, Hermsmeyer (1979).

regarding the mechanisms of vascular smooth muscle activation. An important question is whether or not such quick phasic contractions are present in arterial smooth muscle *in vivo*. Little work has been done in this area. However, the *in situ* recording of spontaneous electrical activity in rat arterials and small arteries (Speeden, 1964; Steedman, 1966) suggests that such a mechanism may play a role in regulating blood flow in the intact animal.

V. SUMMARY

The low E_m in arterial smooth muscle is related to its low resting conductance for K^+. The low g_K and P_K are primarily responsible for the deviation from the Nernst potential for a K^+-selective membrane and the high P_{Na}/P_K ratio seen in arterial smooth muscle cells. However, as is so often the case when studying vascular smooth muscles as a group, it is difficult to make generalizations even regarding resting membrane properties as demonstrated by the high g_K in feline middle cerebral artery. The functional significance of an elevated g_K in some cerebral arteries remains unknown, but may make these vessels more sensitive to the changes in their local ionic environment than other peripheral arteries.

The electrogenic Na^+–K^+ pump provides a sizable electrogenic component to the E_m in arterial muscle and thereby is one factor regulating tension development. This electrogenic component of the E_m varies depending upon $[K]_o$, thereby making determination of the absolute magnitude of its contribution difficult. However, the degree to which the electrogenic pump contributes to the E_m may in part be responsible for the altered reactivity of the arteries observed in certain pathological disease states such as hypertension.

Factors regulating g_K determine to a large extent the excitability of arterial smooth muscle cells. Reduction of the voltage-sensitive outward current by a variety of agents such as TEA allows induction of action potentials in previously quiescent preparations. Stimulation of a Ca^{2+}-sensitive outward K^+ current hyperpolarizes arterial smooth muscle cell membranes, which can provide a stabilizing influence even though intracellular Ca^{2+} is increased. Whether or not such a mechanism exists in the intact animal is yet to be determined. However, the recording of spontaneous electrical activity in arteries *in vivo* suggests that circulatory humoral or neural influences may exist which modify g_K.

The level of E_m in arterial smooth muscle is an important regulator of tension development. Depolarization or hyperpolarization by only 4–6 mV results in significant changes in tension. Many vasoactive agents and neurotransmitters modify tone by changing the level of the E_m. However, it cannot be argued that some arteries respond mechanically to a variety of agents without a change in E_m. In such cases, there are still changes in ion conductance, which can in turn be regulated by the level of the E_m. The membrane response of arterial smooth muscle to neurotransmitters and humoral agents is very much dependent on the location of the artery in the animal and the species of the animal.

The importance of action potential generation in regulating arterial tone is still obscure. The recording of quick phasic electrically mediated contractions in isolated arterial smooth muscle cells suggests that action potential production may provide a mechanism for rapid changes in arterial caliber. However, before such conclusions can be made with any degree of confidence, it will be necessary to further study *in situ* electrical activity of arterial smooth muscle.

ACKNOWLEDGMENTS

The author would like to thank Drs. N. Sperelakis, F. J. Haddy, and K. Hermsmeyer for reviewing this work. The author also wishes to thank F. Diane Dempsey for assistance in preparation of the manuscript. This work was supported by NIH Grant HL-24007.

REFERENCES

Berne, R. M., Belardinelli, L., Harder, D. R., Sperelakis, N., and Rubio, R. (1980). "Response of large and small coronary arteries to adenosine, nitroglycerine, cardiac glycosides, and calcium antagonists." (A. Fleckenstein and H. Roskam, eds.) pp. 208–220. Springer-Verlag, Berlin.

Blaustein, M. P. (1977). Sodium ions, calcium ions, blood pressure regulation, and hypertension: A reassessment and a hypothesis. *Am. J. Physiol.* **232,** C165–C173.

Bohr, D. F. (1973). Vascular smooth muscle updated. *Circ. Res.* **32,** 665–672.

Bolton, T. B. (1979). Mechanisms of action of transmitters and other substances on smooth muscle. *Physiol. Rev.* **59,** 606–718.

Bonaccorsi, A., Hermsmeyer, K., Aprigliano, O., Smith, C. B., and Bohr, D. F. (1977). Mechanism of potassium relaxation of arterial muscle. *Blood Vessels* **14,** 261–276.

Briggs, A. H. (1962). Calcium movements during potassium contraction in rat aortic strips. *Am. J. Physiol.* **203,** 849–852.

Caffrey, J. M., and Anderson, N. C. (1979). Activation of a hyperpolarizing conductance in isolated smooth muscle cells. *Fed. Proc., Fed. Am. Soc. Exp. Biol.* **38,** 972 (abstr.).

Casteels, R., Kitamura, K., Kuriyama, H., and Suzuki, H. (1977a). The membrane properties of the smooth muscle cells of the rabbit main pulmonary artery. *J. Physiol. (London)* **271,** 41–61.

Casteels, R., Kitamura, K., Kuriyama, H., and Suzuki, H. (1977b). Excitation contraction coupling in the smooth muscle cells of the rabbit main pulmonary artery. *J. Physiol. (London)* **271,** 63–79.

Chen, W. T., Brace, R. A., Scott, J. B., Anderson, D. K., and Haddy, F. J. (1972). The mechanism of the vasodilator action of potassium. *Proc. Soc. Exp. Biol. Med.* **140,** 820–824.

Clough, D. H., Pamnani, M. B., Overbeck, H. W., and Haddy, F. J. (1977). Decreased myocardial Na,K-ATPase in rats with one-kidney Goldblott hypertension. *Fed. Proc., Fed. Am. Soc. Exp. Biol.* **36,** 491 (abstr.).

Cuthbert, A. W. (1966). Electrical activity in mammalian veins. *Bibl. Anat.* **8,** 11–15.

Droogmans, G., Raeymackers, L., and Casteels, R. (1977). Electro- and pharmacomechanical coupling in the smooth muscle cells of the rabbit ear artery. *J. Gen. Physiol.* **70,** 129–148.

Dutta, P., Mustafa, S. J., and Jones, A. W. (1980). Effect of adenosine on the uptake and efflux of calcium by coronary arteries of dog. *Fed. Proc., Fed. Am. Soc. Exp. Biol.* **39,** 530 (abstr.).

Funaki, S. (1966). Electrical and mechanical activity of isolated smooth muscle from the portal vein of the rat. *Bibl. Anat.* **8,** 5–10.

Golenhofen, K., and Hermstein, N. (1975). Differentiation of calcium activation mechanisms in vascular smooth muscle by selective suppression with verapamil and D600. *Blood Vessels* **12,** 21–37.

Golenhofen, K., Hermstein, N., and Lammel, E. (1973). Membrane potential and contraction of vascular smooth muscle (portal vein) during application of noradrenaline and high potassium, and selective inhibitory effects of iproveratril (verapamil). *Microvasc. Res.* **5,** 73-80.

Haddy, F. J. (1975). Potassium and blood vessels. *Life Sci.* **16,** 1489-1498.

Haddy, F. J. (1978). The mechanism of potassium vasodilation. *In* "Mechanisms of vasodilation." *Satellite Symp., Int. Congr. Physiol. Sci., 27th, 1978,* pp. 200-205.

Haddy, F. J., and Overbeck, H. W. (1976). The role of humoral factors in volume expanded hypertension. *Life Sci.* **19,** 935-948.

Haddy, F. J., Pamnani, M., and Clough, N. (1978). The sodium-potassium pump in volume expanded hypertension. *Clin. Exp. Hypertens.* **1,** 295-336.

Haeusler, G. (1978). Relationship between noradrenaline-induced depolarization and contraction in vascular smooth muscle. *Blood Vessels* **15,** 46-54.

Haeusler, G., and Thorens, S. (1975). The effects of tetraethylammonium on contraction, membrane potential, and calcium permeability of vascular smooth muscle. *Collog. Inst. Natl. Sante Rech. Med.* **50,** 363-368.

Harder, D. R. (1980a). Comparison of electrical properties of middle cerebral and mesenteric artery in cat. *Am. J. Physiol.* **237,** c23-c26.

Harder, D. R. (1980b). Membrane electrical effects of histamine on vascular smooth muscle of canine coronary artery. *Circ. Res.* **46,** 372-377.

Harder, D. R., and Belardinelli, L. (1980). Effects of Propafenone on TEA-induced action potentials in vascular smooth muscle cells of canine coronary arteries. *Experientia* **36,** 1082-1083.

Harder, D. R., and Sperelakis, N. (1978). Membrane electrical properties of vascular smooth muscle from the guinea pig superior mesenteric artery. *Pfluegers Arch.* **378,** 111-119.

Harder, D. R., and Coulson, P. B. (1979). Estrogen receptors and the effect of estrogens on membrane electrical properties of coronary vascular smooth muscle. *J. Cell Physiol.* **100,** 375-382.

Harder, D. R., and Sperelakis, N. (1979). Action potentials induced in guinea pig arterial smooth muscle by tetraethylammonium. *Am. J. Physiol.* **237,** C75-C80.

Harder, D. R., Belardinelli, L., Sperelakis, N., Rubio, R., and Berne, R. M. (1979). Differential effects of adenosine and nitroglycerine on the action potentials of large and small coronary arteries. *Circ. Res.* **44,** 176-182.

Harder, D. R., Abel, P. D., Hermsmeyer, K. (1981). Membrane electrical mechanism of basilar artery constriction and pial artery dilation by norepinephrine. *Circ. Res.* (in press).

Hermsmeyer, K. (1971). Contraction and membrane activation in several mammalian vascular muscles. *Life Sci.* **10,** 223-234.

Hermsmeyer, K. (1973). Multiple pacemaker sites in spontaneously active vascular muscle. *Circ. Res.* **23,** 244-251.

Hermsmeyer, K. (1976a). Electrogenesis of increased norepinephrine sensitivity of arterial vascular muscle in hypertension. *Circ. Res.* **38,** 362-367.

Hermsmeyer, K. (1976b). Cellular basis for increased sensitivity of vascular smooth muscle in spontaneously hypertensive rats. *Circ. Res.* **38,** Supp II, 53-57.

Hermsmeyer, K. (1976c). Ba^{2+} and K^+ alteration of K^+ conductance in spontaneously active vascular muscle. *Am. J. Physiol.* **320,** 1031-1036.

Hermsmeyer, K. (1979). High shortening velocity of isolated single arterial muscle cells. *Experientia* **35,** 1599-1602.

Hester, R. K., and Carrier, O., Jr. (1976). Excitation contraction/relaxation coupling in vascular smooth muscle. *In* "Factors Influencing Vascular Reactivity" (O. Carrier, Jr. and S. Shibata, eds.), pp. 96-103. Igaku-Shoin, New York.

Hirst, G. D. S., Reild, T. O. (1980). Evidence for two populations of excitatory receptors for noradrenaline on arteriolar smooth muscle. *Nature* **283,** 767-768.

Holman, M. E., and Surprenant, A. M. (1979). Some properties of the excitatory junction potentials recorded from saphenous arteries of rabbits. *J. Physiol.* (*London*) **287,** 337-351.

Holman, M. E., Kasby, C. B., Suthers, M. B., and Wilson, A. F. (1968). Some properties of the smooth muscle of rabbit portal vein. *J. Physiol.* (*London*) **196,** 111-132.

Horn, L. (1978). Electrophysiology of vascular smooth muscle. *Microcirculation* **2,** 119-157.

Isenberg, G. (1977). Cardiac Purkinje fibers: [Ca^{2+}] controls steady state potassium conductance. *Pfluegers Arch.* **371,** 71-76.

Ito, Y., Kuriyama, H., and Suzuki, H. (1978a). The effects of diltiazem (CRDz-401) on the membrane and mechanical properties of vascular smooth muscles of the rabbit. *Br. J. Pharmacol.* **64,** 503-510.

Ito, Y., Suzuki, H., and Kuriyama, H. (1978b). Effects of sodium nitroprusside on smooth muscle cells of rabbit pulmonary artery and portal vein. *J. Pharmacol. Exp. Ther.* **207,** 1022-1031.

Ito, Y., Kitamura, K., and Kuriyama, H. (1979). Effects of acetylcholine and catecholamines on the smooth muscle cell of the porcine coronary artery. *J. Physiol.* (*London*) **294,** 595-611.

Jacobs, A., and Keatinge, W. R. (1974). Effects of procaine and lidocaine on electrical and mechanical activity of smooth muscle of sheep carotid arteries. *Br. J. Pharmacol.* **51,** 405-411.

Johansson, B. (1978a). Vascular smooth muscle biophysics. *Microcirculation* **2,** 83-117.

Johansson, B. (1978b). Processes involved in vascular smooth muscle contraction and relaxation. *Circ. Res.* **43;** *Supp I,* 14-20.

Johansson, B., and Ljung, B. (1967). Sympathetic control of rhythmically active vascular smooth muscle as studied by a nerve-muscle preparation of portal vein. *Acta Physiol. Scand.* **70,** 299-311.

Johansson, B., Johsson, O., Axelsson, J., and Wahlstrom, B. (1979). Electrical and mechanical characteristics of vascular smooth muscle response to norepinephrine and isoproterenol. *Circ. Res.* **21,** 619-633.

Jones, A. W. (1973). Altered ion transport in vascular smooth muscle from spontaneously hypertensive rats. *Circ. Res.* **33,** 563-572.

Kitamura, K., and Kuriyama, H. (1979). Effects of acetylcholine on the smooth muscle cell of isolated main coronary artery of the guinea pig. *J. Physiol.* (*London*) **293,** 119-133.

Kumamoto, M. (1977). Electrophysiological basis for drug action on vascular smooth muscle. *In* "Factors Influencing Vascular Reactivity" (O. Carrier, Jr. and S. Shibata, eds.) pp. 106-130. Igaku-Shoin, New York.

Kuriyama, H., and Suzuki, H. (1978a). Electrical property and chemical sensitivity of vascular smooth muscle in normotensive and spontaneously hypertensive rats. *J. Physiol.* (*London*) **285,** 409-424.

Kuriyama, H., and Suzuki, H. (1978b). The effects of acetylcholine on the membrane and contractile properties of smooth muscle cells of the rabbit superior mesenteric artery. *Br. J. Pharmacol.* **64,** 493-501.

McLean, M. J., and Sperelakis, N. (1977). Electrophysiological recordings from spontaneously contracting reaggregates of cultured vascular smooth muscle cells from chick embryos. *Exp. Cell Res.* **104,** 309-318.

Meech, R. W. (1974). The sensitivity of *Helix aspero* neurons to injected Ca^{2+} ions. *J. Physiol. (London)* **237,** 259-277.
Mekata, F. (1971). Electrophysiological studies of the smooth muscle cell membrane of rabbit common carotid artery. *J. Gen. Physiol.* **57,** 738-751.
Mekata, F. (1974). Current spread in the smooth muscle of the rabbit aorta. *J. Physiol. (London)* **242,** 143-155.
Mekata, F., and Niu, H. (1972). Biophysical effects of adrenaline on the smooth muscle of the rabbit common carotid artery. *J. Gen. Physiol.* **59,** 92-102.
Mulvany, M. J., Halpern, W. (1977). Contractile properties of small arterial resistance vessels in spontaneously hypertensive and normotensive rats. *Circ. Res.* **41,** 19-26.
New, W., Trautwein, W. (1972). The ionic nature of slow inward current and its relation to contraction. *Pfluergers Arch.* **334,** 24-38.
Overbeck, H. W. (1972). Vascular responses to cations, osmolarity, and angiotensin in renal hypertensive dogs. *Am. J. Physiol.* **223,** 1358-1364.
Overbeck, H. W., and Haddy, F. J. (1967). Forelimb vascular response in renal hypertensive dogs. *Physiologist* **10,** 270 (abstr.).
Overbeck, H. W., Pamnani, M. B., Akera, T., Brody, T. M., and Haddy, F. J. (1976). Depressed function of a ouabain-sensitive sodium-potassium pump in blood vessels from renal hypertensive dogs. *Circ. Res.* **381,** Suppl. 2, 48-52.
Pamnani, M. B., Clough, D. L., and Haddy, F. J. (1978). Altered activity of the sodium-potassium pump in arteries of rats with steroid hypertension. *Clin. Sci. Mol. Med.* **55,** 41s-43s.
Schied, C. R., and Fay, F. S. (1980). Control of ion distribution in isolated smooth muscle cells. I. Potassium. *J. Gen. Physiol.* **75,** 163-182.
Schnaar, R. C., and Sparks, H. V. (1972). Response of large and small coronary arteries to nitroglycerine, $NaNO_2$, and adenosine. *Am. J. Physiol.* **223,** 223-228.
Siegel, G., Wiesert, G., Ehehalt, R., Bertsche, O. (1976). Membrane basis of vascular regulation. *In* "Ionic Actions On Vascular Smooth Muscle" (E. Betz, ed.), Springer-Verlag, Berlin, pp. 48-55.
Singer, J. J., and Walsch, J. V. (1980). Penetration-induced hyperpolarization as evidence for Ca^{+} activation of potassium conductance in isolated smooth muscle cells. *Am. J. Physiol.* **239,** C182-C189.
Somlyo, A, V., and Somlyo, A. P. (1968a). Electromechanical and pharmacomechanical coupling in vascular smooth muscle. *J. Pharmacol. Exp. Ther.* **149,** 106-112.
Somlyo, A. V., and Somlyo, A. P. (1968b). Vascular smooth muscle. 1. Normal structure, pathology, biochemistry, and biophysics. *Pharmacol. Rev.* **20,** 197-272.
Somlyo, A. V., Vinall, P., and Somlyo, A. P. (1969). Excitation-contraction coupling and electrical events in two types of vascular smooth muscle. *Microvas. Res.* **1,** 354-373.
Speeden, R. N. (1964). Electrical activity of single smooth muscle cells of the mesenteric artery produced by splanchnic nerve stimulation of the guinea pig. *Nature (London)* **202,** 192-194.
Sperelakis, N. (1980). Origin of the cardiac resting potential. *In* "Handbook of Physiology" (D. T. Bohr, A. P. Somlyo, and H. V. Sparks, eds.), Sect. 2, Vol. I. pp. 187-267. Am. Physiol. Soc., Washington, D. C.
Sperelakis, N., and Shigenobu, K. (1972). Changes in membrane properties of chick embryonic hearts during development. *J. Gen. Physiol.* **60,** 430-453.
Steedman, W. M. (1966). Microelectrode studies on mammalian vascular muscle. *J. Physiol. (London)* **186,** 382-400.
Su, C., Bevan, J. A., and Ursillo, R. C. (1964). Electrical quiescence of pulmonary artery smooth muscle during sympathomimetic stimulation. *Circ. Res.* **15,** 20-27.

Uvelius, B., Sigurdsson, S. B., and Johansson, B. (1974). Strontium and barium as substitutes for calcium on electrical and mechanical activity in rat portal vein. *Blood Vessels* **11,** 245-259.

Van Breemen, C. (1977). Calcium requirement for activation of intact aortic smooth muscle. *J. Physiol.* (*London*) **272,** 317-329.

Van Breemen, C., Aaronson, P., and Loutzenhiser, R. (1979). Sodium-calcium interactions in mammalian smooth muscle. *Pharmacol. Rev.* **30,** 167-208.

von Loh, D. (1971). The effect of adrenergic drugs on spontaneously active vascular smooth muscle studied by long term intracellular recording of membrane potential. *Angiologica* **8,** 144-155.

von Loh, D., and Bohr, D. R. (1973). Membrane potential from smooth muscle cells of isolated resistance vessels. *Proc. Soc. Exp. Biol. Med.* **144,** 513-516.

Waugh, W. H. (1962). Role of calcium in contractile excitation of vascular smooth muscle by epinephrine and potassium. *Circ. Res.* **11,** 927-940.

Webb, R. C., and Bohr, D. F. (1978a). Potassium-induced relaxation as an indicator of Na^+-K^+ ATPase activity in vascular smooth muscle. *Blood Vessels* **15,** 198-207.

Webb, R. C., and Bohr, D. F. (1978b). Mechanism of membrane stabilization by calcium in vascular smooth muscle. *Am. J. Physiol.* **235,** C227-232.

Webb, R. C., and Bohr, D. F. (1979). Potassium relaxation of vascular smooth muscle from spontaneously hypertensive rats. *Blood Vessels* **16,** 71-79.

4

Current Status of Vascular Smooth Muscle Subcellular Calcium Regulation

Julius C. Allen and Richard D. Bukoski

I. INTRODUCTION

The importance of vascular smooth muscle and its role in the regulation of blood pressure and flow has been recognized for many years. A large number of different factors have been studied in attempting to delineate the mechanisms of control of the intact vessel. Factors such as innervation, vessel structure, local humoral effects, and other influences extrinsic to the smooth muscle cell have been deemed critical in proposing mechanisms of the regulation of blood flow to various discrete areas. These have been well characterized and reviewed recently by Vanhoutte (1978) and Altura and Altura (1978). While these authors do not minimize the role of the smooth muscle cell itself, the past has seen significant advances in answering questions as to how smooth muscle cells regulate their contractility. It is now known that calcium interaction

VASCULAR SMOOTH MUSCLE: METABOLIC, IONIC, AND CONTRACTILE MECHANISMS

ISBN 0-12-195220-7

with appropriate myosin-linked receptors initiates contraction much as it does in striated muscle. However, despite the acknowledged role of calcium, little is known about its cytoplasmic regulation. Sarcoplasmic reticulum, sarcolemma, and mitochondria have all been implicated as being critical to the regulation of cellular calcium concentration but, as of now, specific roles for these organelles have not been delineated.

From the previously mentioned work, it is clear that extrinsic factors play a critical role in the regulation of blood flow. In fact, it is felt by some that the well-known observations of heterogeneity in contractile response among different vessels can be explained solely by these extrinsic factors. Little attention has been paid to the concept that the heterogeneous responses of isolated strips of vascular smooth muscle and local regulation in the intact animal could be explained by differences in the mechanisms by which different vascular smooth muscles actually control cytoplasmic calcium.

To date, there has been no unified hypothesis regarding the regulation of cytoplasmic calcium for these tissues. There are a number of possible reasons for this presumed lack of involvement of subcellular calcium regulatory areas in vascular control, but the most critical one is that it has been only since 1972 that subcellular fractions from vascular smooth muscle have been identified as having calcium-sequestering capabilities, despite the fact that the calcium involvement in contraction has been well-known for many years. As a result, workers have suggested that physical or extrinsic factors play a more significant role in regulation than potential differences in calcium-handling characteristics from muscle to muscle.

In 1972 Fitzpatrick *et al.* isolated microsomal fractions and identified their calcium-binding characteristics. Since then, many others have shown that such preparations are capable of sequestering calcium. However, at the present time, there is still considerable controversy as to the specific role of the various membranes contained in these fractions from vascular smooth muscle in regulating calcium and contractility. The present controversy involving the specific roles of sarcolemmal versus sarcoplasmic reticulum binding is indicative of the confused state regarding vascular smooth muscle. The immediate source of activator calcium is not known, nor is the mechanism of relaxation known, i.e., where the calcium comes from and where it goes.

There are many other areas external to the cell that may serve to regulate contractility in the whole vessel. Their description need not be repeated here, but, by this omission, it is not intended to minimize their importance. Rather, it is the purpose of this chapter to suggest that subcellular regulatory characteristics play an important role in the regulation of vascular smooth muscle contractility and that differences in these regulatory characteristics among vessels may contribute to vascular smooth muscle heterogeneity. The fact of the matter is that available data do not yet support such a contention. Data are, however, now becoming available for use in an attempt to understand the important roles of different subcellular areas of vascular smooth muscle tissue, and methodologies are becoming available to identify their calcium regulatory roles.

We would like, at the outset, to forward the hypothesis that in some instances vascular smooth muscle heterogeneity can be the result of differences in subcellular handling of calcium. Such a suggestion was made earlier by virtue of observations of ultrastructural differences in the vesicle content among different intact vascular smooth muscles (Devine *et al.,* 1973). This example not withstanding, the data are not yet available to support the presented hypothesis. However, present and current physiological evidence for vascular smooth muscle heterogeneity from the literature will be presented and how this might be explained by differences in the handling of calcium will be suggested. Various biochemical characteristics of subcellular areas with respect to calcium metabolism and its regulation will be systematically identified, and the type of data that must be obtained in order to understand the role or roles that a specific organelle may play will be specified. It is hoped that it will be clarified why contractile heterogeneity has not yet been explained on a subcellular basis. It is not because such a basis does not exist, but that the proper questions have not been asked nor properly answered. Finally, because of the importance of subcellular organelles in the regulation of intracellular calcium concentration, current concepts of organelle involvement as key components of cyclic adenosine 3′, 5′-monophosphate (cAMP)-mediated events and in the development and maintenance of hypertension will be briefly discussed.

II. PHYSIOLOGICAL HETEROGENEITY

It has become apparent that the study of subcellular calcium handling in vascular smooth muscle can be and has been used in attempts to explain (1) differences in vascular smooth muscle from hypertensive and normotensive animals, (2) differences in the mechanism of action of various pharmacological agents on a single blood vessel, and (3) differences in blood vessels displaying heterogeneous responses to the same pharmacological agents.

A major effort has been directed toward the use of subcellular calcium-binding data in the study of potential differences between normotensive and hypertensive blood vessels. For this reason, a separate section of this chapter (Section V) has been devoted to this area. The use of subcellular tissue studies in the examination of pharmacological heterogeneity within a single vessel and among different vessels has received comparatively less attention. As outlined in Section I, we will present a number of examples of heterogeneity which can perhaps be explained by differences in calcium handling, without reiterating the previous comprehensive reviews already mentioned. The current status of the subcellular biochemical processes possibly involved in calcium metabolism will then be presented. We hope that it will become as apparent to the reader as it has to us that specific knowledge of these subcellular processes is quite limited.

The main body of work examing physiological heterogeneity has been primarily concerned with the heterogeneous responses of a given vessel to a number of different agonists. Examples of responses involving calcium are presented in extensive reviews by Weiss (1977) and Altura and Altura (1978). For this reason, examples of heterogeneity among different vessels in response to a single agonist will be stressed here. The literature shows that there is a quantitative difference in response among rat aorta, femoral artery, and portal vein, both to the alpha-adrenergic agonists, epinephrine and norepinephrine, and to potassium ions. The maximal normalized mechanical responses (in grams per square centimeter) to both types of agonists differed in the following order: femoral artery (Hansen *et al.,* 1974) > aorta (Shibata *et al.,* 1973) > portal vein (Greenberg and Bohr, 1975). On the assumption that differences in actomyosin content and activity per cross-sectional area among the vessels cannot totally explain these differences (Murphy, 1979), it can be suggested that part of the reason for the variable responses is a difference in the dynamic nature of calicum metabolism. Perhaps there are release sites that are more active than uptake sites in muscles displaying greater tension. There are certainly no data suggesting that release and uptake sites (or perhaps their activities) are similar, but they may in fact exhibit differing temporal latencies. In general, these sites have not been defined in any vascular smooth muscle. Calcium pools exist, but knowledge of their location and identification with specific organelles is lacking (Kutsky *et al.,* 1980).

Bohr *et al.* (1971), using helical strips of aortas and small arteries of the rabbit, observed differential responses to caffeine and calcium depletion. They suggested that the reason for the difference in results was the existence of different calcium pools in the different vessels, but further definition of these pools was limited.

Somlyo and Somlyo (1968, 1970) classified vascular smooth muscle into spike-generating (phasic) and graded responsive (tonic) types.

Golenhofen (1976) has hypothesized a model of phasic and tonic systems (P and T) functional in these two types of muscle that handle calcium differently and are responsible for the two types of responses.

Hanley *et al.* (1975) determined the effects of the ionophore RO 2-2985 in intact, chronically instrumented dogs. They observed a heterogeneous effect on coronary, renal, and iliac artery blood flow. Subsequently, our laboratory (Bukoski *et al.*, 1979) observed heterogeneous effects of the same ionophore on *in vitro* muscle preparations of canine coronary, renal, and femoral arteries. Specifically, RO 2-2985 relaxed KC1-contracted coronary arteries but had no effect on KC1-contracted renal or femoral arteries. The calcium-specific ionophore A23187 had little effect on any of the arteries. RO 2-2985 also had a relaxing effect on phenylephrine-contracted renal and femoral arteries, but A23187 had no effect. RO 2-2985 is known to transport calcium, as well as other monovalent and divalent cations, and it is conceivable that the differential responses are the result of different calcium pools or calcium-handling properties of the vessels.

In a similar study using neomycin rather than RO 2-2985, Adams and Goodman (1975) demonstrated a differential effect of the antibiotic on a number of canine vessels. They concluded that the heterogeneous responses were due to differing calcium dependencies of the various arteries.

From these few examples, it is clear that there is evidence suggesting that differences in calcium-handling characteristics can result in contractile heterogeneity among vascular smooth muscles from different anatomical locations.

III. SUBCELLULAR CALCIUM REGULATION: BIOCHEMICAL APPROACHES

The experimental approaches that have been made will now be examined in order to understand calcium handling on the molecular and subcellular levels. It will then be determined how close we are to answering the question as to whether or not differences in subcellular calcium regulation contribute to the mechanical differences observed in heterogeneous responses and in hypertension.

A. Mitochondria

A fundamental problem in identifying the specific sites of calcium control in vascular smooth muscle has been the fact that it has been difficult to isolate either pure or enriched fractions of sarcoplasmic reticulum, sarcolemma , and mitochondria from this tissue. Because of these technical difficulties and the very heterogeneous nature of the isolated fractions, a calcium-sequestering capability has been identified with all three organelles. Recently, a number of workers has successfully isolated and identified mitochondrial fractions of sufficient purity, so that some specific characteristics regarding calcium uptake could be identified (Vallieres *et al.*, 1975; MacNamara *et al.*, 1979). Most of the previous data regarding calcium binding by mitochondrial preparations of vascular smooth muscle were demonstrated using ATP as the high-energy substrate, rather than respiratory substrates such as malate and succinate. The choice of using ATP or a respiratory substrate as the energizing compound is difficult to make, since both will support calcium uptake. Respiratory substrates, however, may be more closely related to *in vivo* situations. Mitochondria, using appropriate substrates, produce a high-energy state in the form of a proton gradient across their inner and outer membranes (Boyer *et al.*, 1977). Oxidative phosphorylation and calcium uptake may compete with each other for this transient high-energy state. If ADP is present, the mitochondria will utilize this gradient to produce ATP. However, if ADP is absent, the potential energy can then be used for one of many of the other mitochondrial functions, such as calcium transport (Lehninger *et al.*, 1978).

The various respiratory parameters classically used to assess tightness of coupling and purity of mitochondrial preparations had not generally been studied in preparations from vascular smooth muscle until the work of Vallieres *et al.* (1975). These workers presented appropriate assessment data and further examined the mitochondrial calcium sequestration capacities of bovine main pulmonary artery and mesenteric vein. They concluded that these organelles could take up enough calcium to perhaps be involved in excitation–contraction coupling in these tissues. The K_m of the calcium uptake was 17 μM, and the rate of uptake was measured to be in excess of 10 nmol mg^{-1} sec^{-1}. It has been shown, however, that for an 80% maximal contraction of vascular smooth muscle, the free calcium concentration is assumed to be approximately 1 μM (Sloane, 1980). Because of the K_m of 17 μM, therefore, there would be very little mitochondrial calcium transport at this relatively low concentration. However, Devine *et al.* (1973) had earlier demonstrated that there appeared to be a large quantity of mitochondria in mesenteric vein smooth muscle cells from the rabbit. If indeed mitochondria actually sequester a high level of calcium without uncoupling oxidative phosphorylation (Vallieres *et al.* found 210 nmol/mg of mitochondrial protein), they would generate a significant calcium sink for this tissue. It can be suggested that mitochondrial calcium-sequestering capacity or sensitivity may be considerably altered by altered ionic strength or phosphorylation parameters. Thus, even though mitochondrial involvement is unlikely, total disregard at this point would be premature. In addition, thus far there has been no suggestion that physiological parameters can effect the release of this sequestered calcium in mitochondria, as has been done for cardiac muscle by Crompton *et al.* (1978). These workers showed that sodium could effect a specific release of calcium, which had been taken up by mitochondria appropriately energized by respiratory substrates. These data led them to postulate a sodium-dependent efflux of calcium from cardiac mitochondria. The calcium release phenomenon is a critical parameter to be studied in suggesting that any organelle is involved in excitation–contraction coupling, since the sequestration of calcium is only one-half of the excitation–contraction relaxation cycle.

It also should be mentioned here that these cardiac mitochondria have been shown to sequester calcium quite readily, but whether this is of physiological significance is still undecided (Scarpa, 1976). Most workers feel that participation in beat-to-beat control seems unlikely, since the rates and quantities of calcium binding are too slow and too small to be involved in a physiological situation. Also, cardiac sarcoplasmic reticulum and sarcolemmal vesicles have a far greater capacity for calcium uptake than cardiac mitochondria. At present, then, even though smooth muscle mitochondria sequester calcium and exist in a relatively large number, it is questionable whether or not these organelles actually participate in physiological regulation of vascular smooth muscle contractility.

Sloane *et al.* (1978) showed some extremely interesting magnesium fluctuations in mitochondria isolated from bovine main pulmonary arteries. These mitochondria exhibited specific, respiratory substrate-supported release of magnesium. This was not unlike the magnesium efflux observed from cardiac mitochondria by Crompton *et al.* (1976), but the rate in the cardiac preparation was much less than that observed with vascular smooth muscle mitochondria. Even though the total free magnesium concentration may be decreased by as much as 0.2 m*M* by such mitochondrial sequestration of magnesium, the significance of this characteristic of vascular smooth muscle mitochondria is unknown at present.

In conclusion, we would like to question the value of studying calcium sequestration by mitochondria supported by ATP rather than respiratory substrates, especially in the absence of appropriate assessment of respiratory parameters. It can be argued that data accumulated under such conditions may have little or no relationship to *in vivo* physiological processes, and that future investigations into the role of vascular smooth muscle mitochondria in calcium sequestration be carried out with these considerations in mind. It seems unlikely at the moment that these organelles are involved in specific regulation of vascular smooth muscle contractility. They may serve as a sink, perhaps in pathological situations, but it seems improbable that they are involved in tonic or phasic regulation.

B. Sarcolemma and Sarcoplasmic Reticulum

1. Subcellular Fractionation

The major biochemical emphasis for the past several years in attempts to understand and identify specific regulatory mechanisms in vascular smooth muscle has been the study of sarcoplasmic reticulum and sarcolemmal fragments. A number of laboratories have been successful in the isolation of microsomal fractions from vascular smooth muscle presumed to contain chiefly sarcoplasmic reticulum that is capable of sequestering calcium on addition of ATP (Table I). It has been suggested that these particles are the counterparts of sarcoplasmic reticulum in both cardiac and skeletal muscle. However, a majority of these studies have suffered from the same shortcoming: difficulty in separation and identification of the presence of sarcolemmal or sarcoplasmic reticular membrane fragments. Although enzyme markers have been used in attempts to identify the various subcellular organelles, valid conclusions reached regarding the specific membrane content of subcellular fractions have been very incomplete. As a result, it has been difficult to specify the membrane to which calcium binding occurs. The large majority of studies to be discussed isolated microsomal fractions, and in only a few were enzyme markers identified. Among these were 5′-nucleotidase, adenyl cyclase, K^+-phosphatase, and alkaline phosphatase for sarcolemma, NADH-dependent, rotenone-insensitive cytochrome *c* reductase and ATP plus oxalate-stimulated calcium uptake for sarcoplasmic reticulum, and cytochrome *c* oxidase and succinate dehydrogenase for mitochondria. The general assumption has been that 5′-nucleotidase is sufficient for identification of the presence of sarcolemma-like material. Also, there is no question that the microsomal fractions studied sequester calcium in an oxalate-stimulated manner, but the specific membranes involved in this process have not been indisputably identified.

Some years ago, Hurwitz and associates (1973) attempted to correlate the biochemistry of subcellular fractions with the ability of different muscle types to contract in the absence of external calcium. They iso-

TABLE I

Calcium Uptake and ATPase Activities in Membrane Preparations[a]

Vessel	Fraction[b]	Method[c]	Ca Sequestration			Ca^{2+}-ATPase	Markers[d]	Reference
			ATP	Oxalate	[Ca]			
Rabbit aorta	M	1	20 nmol/mg · 1 min	60 nmol/mg · 1 min	100 μ*M*	0.4 nmol P/mg · 20 min; [Ca], 20 μ*M*	5′ and SDH	Fitzpatrick *et al.*, 1972
	Mt	1	25 nmol/mg · 1 min	—	100 μ*M*		5′ and SDH	
Rabbit aorta	M	1, 2	—	80 nmol/mg 10 min	60 n*M*	—	5′, NKA, NADH ox.	Hurwitz *et al.*, 1973
Rabbit aorta	M	1	6.5 nmol/mg · 10 min	—	100 μ*M*	14 nmol μP/mg · 10 min; [Ca], 2 n*M*	Cyt *c* ox.	Baudouin and Meyer, 1972
Rabbit aorta	M	1, 2	Data not reported			Data not reported	Cyt *c* ox., 5′, G-6-P′tase, AC, Mg^{2+}-ATPase	Devynck *et al.*, 1973
Bovine aorta	M	1	17 nmol/mg · 8 min	40 nmol/mg · 8 min	45 μ*M*	2.5 μmol P/mg · 1 hr	Cyt *c* ox., 5′, NKA	Hess and Ford, 1974
	Mt	1	75 nmol/mg · 8 min	70 nmol/mg · 8 min	10 μ*M*	—	Cyt *c* ox., 5′, NKA	
Hog coronary artery	M	1	7.5 nmol/mg · 10 min	9 nmol/mg · 10 min	100 μ*M*	—	—	Zelck *et al.*, 1975
	Mt	1	9 nmol/mg · 10 min	—	100 μ*M*	—	—	
Bovine aorta	M	1	18 nmol/mg · 8 min	55 nmol/mg · 8 min	45 μ*M*	—	—	Ford and Hess, 1975

(*continued*)

TABLE I (*Continued*)

Vessel	Fraction[b]	Method[c]	Ca Sequestration: ATP	Ca Sequestration: Oxalate	Ca Sequestration: [Ca]	Ca^{2+}-ATPase	Markers[d]	Reference
Rat mesenteric artery	PM	1, 2	15.3 nmol/mg · 10 min	16 nmol/mg · 10 min	100 μM	—	5′, K-P′-tase, alk. P′tase, cyt *c* ox., PDE I	Wei *et al.*, 1976a
	SR	1, 2	6.9 nmol/mg · 10 min	15 nmol/mg · 10 min	100 μM	—		
Hog coronary artery	M	1, 2	16 nmol/mg · 10 min	30 nmol/mg · 10 min	20 μM	—	5′, NADH red., CPPT, and AC	Wuytack *et al.*, 1978
Rabbit mesenteric artery	M	1, 2	4.8 nmol/mg · 10 min	12 nmol/mg · 10 min	1 μM	2.5 μmol P/mg · 1 min; [Ca], 5 mM	—	Thorens, 1979
Hog coronary artery	M	1, 2	20 nmol/mg · 10 min	105 nmol/mg · 10 min	10 μM	20 μmol P/mg · 10 min	—	Wuytack and Casteels, 1980

[a] Tables I, III, and IV are designed to show the reader the varied nature of separate attempts to study similar phenomena. Despite the fact that similar isolation procedures are used, note the use of different vessels and markers. No attempt has been made to be inclusive in expressing calcium sequestration, ATPase data, or cAMP affects. We have tried to be selective for comparison. Data are from the indicated publications.

[b] M, Microsomes; Mt, mitochondria; PM, plasma membrane; SR, sarcoplasmic reticulum.

[c] 1, Differential centrifugation; 2, sucrose density gradient.

[d] 5′, 5′-Nucleotidase; SDH, succinic dehydrogenase; NKA, Na^+, K^+-ATPase; NADH ox., NADH oxidase; cyt *c* ox., cytochrome *c* oxidase; G-6-P'tase, glucose-6-phosphatase; AC, adenyl cyclase; K-P'tase, K phosphatase; alk. P'tase, alkaline phosphatase; PDE I, phosphodiesterase I; NADH red., NADH reductase; CPPT, choline phosphotransferase.

lated microsomal fractions from guinea pig longitudinal smooth muscle and rabbit aorta and layered them on sucrose density gradients. In longitudinal muscle, the results suggested that calcium uptake corresponded to specific fractions also containing Na^+,K^+-ATPase activity, generally accepted to be a marker for cell membrane. Aortic material possessed calcium uptake capability in a fraction which did not contain cell membrane. In this study 5′-nucleotidase was used as the cell membrane marker, since at that time it was suggested that Na^+,K^+-ATPase was not found in rabbit aorta. However, both our laboratory (Allen and Seidel, 1978) and others (Wolowyk *et al.*, 1971) have identified Na^+,K^+-ATPase from rabbit aorta. One must seriously question the validity of using only 5′-nucleotidase as a marker for sarcolemma in the absence of Na^+,K^+-ATPase activity. The conclusions of Hurwitz's group were based on physiological responses of rabbit aorta and guinea pig ileum observed in a calcium-free medium. It had been found that norepinephrine-induced contractions of rabbit aorta did not require extracellular calcium but utilized intracellular calcium, and a correlation with the biochemical data suggested that the calcium rate-limiting step would be found in the sarcolemmal calcium system as its rate-limiting site.

Probably the most extensive attempt at purifying, characterizing, and separating sarcolemma from sarcoplasmic reticulum in vascular smooth muscle has been the work of Daniel and his colleagues (Wolowyk *et al.*, 1971; Wei *et al.*, 1976a,b,c; 1977a,b; Kwan *et al.*, 1979a,b; Kwan and Daniel, 1980). They extensively characterized their preparations of subcellular membranes using NADPH cytochrome *c* reductase as an endoplasmic reticulum marker and 5′-nucleotidase and potassium-stimulated phosphatase as sarcolemmal markers. Even though these preparations were significantly cleaner and more enriched than others, Na^+,K^+-ATPase was not systematically characterized. In only two of the preparations of Wolowyk *et al.* (1971) and Wei *et al.* (1976a), using rabbit aorta and mesenteric artery, respectively, was Na^+,K^+-ATPase listed. In the earlier study, Na^+,K^+-ATPase was found to be stimulated by a soluble factor, whereas in the later study, the Na^+,K^+-ATPase activity needed no such factor but was still low, about 15 μmol P_i mg^{-1} hr^{-1} in the presence of a Mg^{2+}-ATPase activity of over 350 μmol P_i/mg protein per hour. If Na^+,K^+-ATPase is required in vascular smooth muscle for the maintenance of an ionic gradient, as in other tissues (Fleming, 1980; Jones, 1980), why is the presence of this enzyme not more readily demonstrable in vascular smooth muscle preparations?

2. Na^+, K^+-*ATPase*

Although in recent years the existence of Na^+,K^+-ATPase has been demonstrated in vascular smooth muscle, there have been no preparations upon which kinetic studies, such as ouabain binding and inhibition and the determination of various activator K_m values, have been performed. As indicated earlier, published enzyme activities are wide-ranging for Na^+,K^+-ATPase, and the activity has generally been obtained in the presence of a large quantity of Mg^{2+}-ATPase activity. The overwhelming presence of this latter nonspecific activity makes further specific examination of Na^+,K^+-ATPase quite difficult, since it may comprise less than 10% of the total activity. There is also another problem in that the simple appearance and measurement of Na^+,K^+-ATPase activity are not sufficient to elucidate the role that the sarcolemmal surface may play in regulating calcium flux. There have, however, been some attempts to overcome these difficulties.

A recent study of Preiss and Banaschak (1979) has attempted to correlate Na^+,K^+-ATPase activity with calcium regulation in bovine common carotid artery. They examined the effects of ouabain and changes in sodium and potassium concentrations on a number of cellular and subcellular functions, including whole-tissue ^{45}Ca efflux, Na^+,K^+-ATPase and Ca^{2+}-ATPase activities, and microsomal calcium uptake. They observed some correlation of the effect of increased and decreased potassium concentration and ouabain addition on the Na^+,K^+-ATPase and ^{45}Ca efflux parameters. They concluded that the activity of the sodium pump may regulate the tone of large arteries through alterations in calcium storage processes.

Na^+, K^+-ATPase activities of vascular smooth muscle preparations are at least an order of magnitude lower than those of similar preparations of cardiac and skeletal muscle. While this may be due to technical difficulties in enzyme isolation, the low activity may in fact reflect real biological differences in the number of pump sites available per unit membrane area. For example, Brading and Widdicombe (1974) have calculated that, for guinea pig taenia coli, there are approximately 250–300 mol of ouabain-binding sites per square micrometer of membrane or 3 $\times$ 10^6 mol of ouabain-binding sites per square centimeter. This figure is approximately three orders of magnitude less than comparable calculations made for cardiac muscle, which yield a sodium pump site density of approximately 6 $\times$ 10^9 mol/cm^2 (Kuschinsky *et al.*, 1968; Michael *et al.*, 1979). If then the number of sodium pump sites per unit smooth muscle cell membrane area is significantly less than in cardiac muscle, it is not surprising that a reasonable membrane preparation from smooth muscle may have a much lower Na^+,K^+-ATPase activity than a comparable cardiac preparation.

Recent data from this laboratory compared [^{3}H]ouabain binding and Na^+,K^+-ATPase activity from guinea pig vas deferens and heart (Gerthoffer and Allen, 1981). It was suggested that Na^+,K^+-ATPase activity of heart was approximately 500 units per gram wet weight, whereas the vas deferens contained 13.5 units. The heart homogenate contained approximately 517 pmol [^{3}H]ouabain binding per gram wet weight of tissue, whereas the guinea pig vas deferens crude membrane contained only 75 pmol [^{3}H]ouabain bound per gram wet weight. Therefore, there was a difference of about six times the total tissue-binding sites in the heart compared to the vas deferens. Such data suggest that isolation of a membrane fraction of vas deferens comparable to heart would have significantly lower Na^+,K^+-ATPase activity than the heart. Naturally, one must assume availability of all ATPase and binding sites. While this is not conclusive evidence of a different number of ouabain-binding sites in smooth and cardiac muscle, it suggests that ouabain-binding sites per unit membrane area may be different and lends credence to the suggestion that low enzyme activity from smooth muscle is not simply a technical problem.

The suggestion of different numbers of Na^+,K^+-ATPase sites per unit cell membrane is also consistent with the ultrastructural comparison of vascular smooth muscle and cardiac muscle. The vascular smooth muscle membrane surface contains significantly more membrane vesicles than the cardiac membrane surface. Therefore, there could conceivably be larger proportions of cell membrane concerned with functions other than sodium and potassium transport in vascular smooth muscle compared to cardiac muscle. The synthetic capacity and attending secretory characteristics of vascular smooth muscle are well known (Ross and Kariya, 1980). Although these suggestions are speculative, the data suggest that there may be differences in the activity obtainable from these preparations, and the problem may not simply be technical in nature.

3. Calcium Sequestration Studies

The calcium uptake capacity of sarcoplasmic reticulum has been shown to be stimulated by oxalate, a permeable anion. Daniel's group has suggested that sarcoplasmic reticulum requires less oxalate for stimulation than sarcolemma. It is suggested then that oxalate does not readily penetrate vesicles of sarcolemmal origin but can penetrate those of sarcoplasmic reticulum. In addition, these differing permeabilities and sensitivities to oxalate stimulation do not hold for another permeant anion, phosphate. This difference in sensitivity to precipitating anions may mirror actual vesicle permeability differences or differences in vesicle sidedness in the two membrane preparations (Kwan *et al.*, 1980). Similar selective membrane differences have been helpful in attempting to separate these membranes in cardiac muscle preparations. A mixture of membrane vesicles is loaded with oxalate and then placed on 30% sucrose. Upon centrifugation, the denser oxalate-containing vesicles (sarcoplasmic reticulum) migrate to the bottom of the centrifuge tube, whereas vesicles not loaded with oxalate (sarcolemma) remain localized at the top (Jones *et al.,* 1979). Unfortunately, our laboratory's attempt to repeat this procedure for canine aortic vesicles has not met with success.

Another group that has studied calcium binding and uptake in vascular smooth muscle is Bhalla and co-workers, using largely rat aorta (Webb and Bhalla, 1976a,b; Bhalla *et al.,* 1978a,b). A major problem with their work, however, is that they were not able to separate sarcoplasmic reticulum and sarcolemmal material. They used a 100,000 *g* pellet, microsomes that contained in all probability both sarcoplasmic reticulum as well as sarcolemma. The marker used for the presence of sarcolemma was 5′-nucleotidase. They did not assay for Na^+,K^+-ATPase. In a similar rat aortic preparation, we were also unable to demonstrate the presence of Na^+,K^+-ATPase. However, when we treated these preparations with sodium dodecyl sulfate (SDS) for 20 min at room temperature, a ouabain-inhibited Na^+,K^+-ATPase activity was observed (Allen and Seidel, 1978). So, despite the fact that these workers were unable to demonstrate the presence of any sarcolemmal material in their endoplasmic reticulum fraction, some of the calcium binding measured may have been due to material of cell membrane origin. Detergent treatment for revealing latent Na^+,K^+-ATPase activity has also been used in cardiac preparations (Besch *et al.,* 1976).

Another factor also enters into significance when one considers the following example: If calcium uptake is determined in a reaction volume of 1 ml containing 2 m*M* ATP, and a Mg-ATPase activity of 50 μmol P_i/mg protein per hour is present, then in 10 min the Mg^{2+}-ATPase alone will reduce the ATP content from 2 μmol ATP to 1.2 μmol ATP, a reduction of 42%. One wonders what effect these changes in ATP content have upon calcium binding occurring over a 30- to 60-min period. This becomes particularly important when one compares calcium binding to these microsomal fractions from hypertensive and normotensive animals (see Section V). The differences seen in calcium binding may simply mirror differences in Mg^{2+}-ATPase activity. Consequently, calcium-binding experiments performed without regard for maintenance of ATP concentrations over the entire course of the experiment may not be legitimate for use in identifying differences in pathological conditions. In addition to these considerations, when perusing the literature on calcium uptake in subcellular fractions, it quickly becomes evident that the experimental conditions, i.e., [ATP], free [Ca^{2+}], pH, and length of incubation vary widely, and this prevents the easy comparison of results among species, tissues, and laboratories (Table I).

Wuytack and his colleagues (1978) have also studied the calcium ATP-dependent transport system of microsomal fractions from pig coronary artery. Again, they did not show significant Na^+,K^+-ATPase activity in any of these fractions and used only 5′-nucleotidase as their sarcolemmal enzyme marker. They showed oxalate-dependent calcium uptake requiring ATP, and on further density gradient centrifugation of this fraction, demonstrated considerable heterogeneity of calcium uptake capability.

Table III of their paper clearly shows that there was an enrichment of 5′-nucleotidase and adenyl cyclase activity along with an enrichment of oxalate-stimulated calcium accumulation in the microsomal fraction. Subsequent density gradient centrifugation then provides a partial separation of these membranes. However, at certain gradient levels, if oxalate stimulation is used as a criterion for the presence of sarcoplasmic reticulum, there appears to be contamination of the cell membrane fraction with sarcoplasmic reticulum and the membrane separation is not complete (see Fig. 3 of that paper).

A later study by Wuytack and Casteels (1980) attempted to show that the Ca^{2+} and Mg^{2+}-ATPase activity demonstrable in a microsomal fraction of pig coronary artery was likely to be involved in calcium transport. This conclusion was based on some similarities in the kinetic characteristics of calcium and magnesium activation of ATPase activity and those of calcium uptake. The basal Mg^{2+}-ATPase activity indicated in Table II of their paper is significantly lower than other published activities. The activity of 32.99 nmol P_i/mg protein per minute (1.98 μmol P_i/mg protein per hour) is significantly lower than that for most other vascular smooth muscle preparations. If the activity of this enzyme is in actuality so much lower than that found in other tissues, the amount of ATP hydrolysis over a 30-min period would not become a limiting factor in the calcium uptake studies, but it might also indicate that the membrane preparation was substandard.

4. *Separation of Sarcolemma and Sarcoplasmic Reticulum*

It is critical to be able to identify positively the presence of both sarcoplasmic reticulum and sarcolemma with appropriate marker enzymes, especially in view of the lack of consistent identification of Na^+,K^+-ATPase activity. [^{3}H]ouabain binding certainly can be used as a positive marker for the presence of sarcolemma. However, there is no specific evidence that 5′-nucleotidase is a specific sarcolemmal marker for vascular smooth muscle. Indeed, it has not yet been correlated with Na^+,K^+-ATPase activity in this tissue.

Daniel and his co-workers have used potassium-stimulated, ouabain-inhibited phosphatase activity as a membrane marker. This activity copurifies with Na^+,K^+-ATPase activity in most other tissues, and this fact has been used to justify its use as a marker. However, since it is presumed to represent the second half of the overall Na^+,K^+-ATPase reaction, it may only indicate the presence of membrane fragments rather than intact vesicles, in which case such fragments probably could not sequester calcium. In addition, such expression of only partial reactions may reflect vesicle heterogeniety, with respect to sidedness and intactness. All marker systems in use then have partial basis in fact, but all are recognized as substitutes for the measurement of Na^+,K^+-ATPase. Thus far we have determined that membrane fractions, be they sarcolemmal or sarcoplasmic reticulum, sequester calcium and may be regulatory in their nature in vascular smooth muscle. The major question to be asked, however, is, How can the two membrane fractions, sarcolemma and sarcoplasmic reticulum, be separated? The density gradient centrifugation studies of Hurwitz, Wuytack, and Daniel and their co-workers, as well as those of our own laboratory, have indicated that there is extraordinary heterogeneity in all these gradient fractions, despite the fact that there are significant increases in the ATPase activity. Placement of microsomal material on density gradients in our laboratory has given Na^+,K^+-ATPase activity up to 20–30 μmol P_i/mg protein per hour, but this occurs in the face of large quantities of Mg-ATPase activity (Table II). Sarcolemmal material must be separated from other membrane material in order to determine whether or not sarcolemma has a calcium-binding capacity. The work of Wei *et al.* (1976a,b, 1977a,b) and Kwan *et al.* (1979a,b, 1980) effected a partial separation of membrane material, but it did not appear to have a very active Na^+,K^+-ATPase. This is of considerable importance, if one ultimately must answer the question as to whether Na^+,K^+-ATPase plays a role in calcium-binding capabilities (Preiss and Banaschak, 1979).

We have placed microsomal suspensions on discontinuous sucrose gradients and assayed Na^+,K^+-ATPase as a sarcolemmal marker. The data (Table II) show that the membranes isolated at the 30 and 40% interfaces, although of different density, have essentially the same Mg^{2+}-ATPase and Na^+,K^+-ATPase activities. These data are consistent with the other density gradient studies of Daniel's group (Wolowyk *et al.*, 1971; Wei *et al.*, 1976a,b,c; Kwan *et al.*, 1979a) and Wuytack *et al.* (1978)

TABLE II

ATPase Activity of Canine Aortic Membrane Fractions[a]

	30% Sucrose gradient	40% Sucrose gradient	50% Sucrose gradient
Mg^{2+}	132.6	114.4	51.3
Mg^{2+} + Na^+ + K^+	174.8	155.2	61.9
Mg^{2+} + Na^+ + K^+ ouabain	154.2	139.8	60.8
Ouabain-inhibitable Na^+, K^+-ATPase	20.6	15.4	1.1

[a] The assay conditions were as follows: imidazole, 50 m*M*, pH 7.4; Mg, 5 m*M*; Tris-ATP, 5 m*M*; Na, 100 m*M*; K, 10 m*M*; ouabain, 100 μ*M*. The units of enzyme activity are micromoles of Pi per milligram of protein per hour. Note the distribution of Na^+, K^+-ATPase activity in the 30 and 40% fractions.

in that similar activities are found in fractions of significantly different density. Hess and Ford (1974), using bovine aorta, made a similar observation in that Na^+,K^+-ATPase was enriched in both mitochondrial and microsomal fractions. They suggested that the reason for the diffuse appearance of the enzyme was that it was not tightly bound to the cell membrane. It is conceivable that the reason for the appearance of Na^+,K^+-ATPase at different densities in sucrose gradient preparations is that Na^+,K^+-ATPase is loosely bound to the membrane, but this is unlikely, since Na^+,K^+-ATPase is believed to span the membrane and therefore probably could not be easily disrupted (Schwartz *et al.*, 1975).

Another possible explanation for such data, which differ from the gradient results obtained with cardiac or skeletal muscle, may be found in an examination of the recent paper of Forbes *et al.* (1979). It is well known that the ultrastructure of vascular smooth muscle is not nearly as regular as that of cardiac or skeletal muscle and that there appear to be considerable areas of sarcolemma comprised of surface vesicles, or caveolae. In addition, the contractile protein array in these tissues is more varied in its orientation than in other muscle tissue. It is possible that, because of this extensive tissue-to-tissue variability, there is an ultrastructural definition applicable to heterogeneity and calcium pools. For further discussion of these possibilities the reader is referred to A. V. Somlyo (1980).

Forbes *et al.* have shown significant variation in the nature and appearance of caveolae. Their micrographs depict a significant branching and beading effect of these vesicles, so that groups of them may project deep into the cell interior. In addition, it is known that sarcoplasmic reticulum vesicles in smooth muscle are very irregular in nature. Forbes's group again has shown that these organelles can traverse almost the entire cell, beginning deep inside and terminating as junctional sarcoplasmic reticulum at the membrane surface. Because of this very irregular intracellular organization, it is suggested that cellular disruption may also follow irregular paths. When cardiac or skeletal muscle cells are physically disrupted by homogenization, it seems logical, because of the regular internal array, that subcellular membranes (sarcoplasmic reticulum and sarcolemma) may then be relatively easily separated by simple centrifugation. By placing the partially separated fraction on a sucrose gradient, a further separation would result. Unlike the situation in striated muscle, however, disruption of vascular smooth muscle may effect an irregular separation, so that a given unit of homogenate would contain a variety of membrane material of similar density. Thus a microsomal fraction from vascular smooth muscle would be far more heterogeneous than a similar fraction from cardiac or skeletal muscle. Furthermore, separation of smooth muscle fractions by density gradient would yield bands of considerable density difference but with a very similar mixed membrane content. As a result of these factors, techniques that have successfully separated these membranes in striated muscle may not be appropriate for achieving such a separation in vascular smooth muscle.

One type of investigation which may aid in solving this problem is the study of lipid/protein ratios and phospholipid analysis of vascular smooth muscle membranes, the assumption being that cell membranes and sarcoplasmic reticulum contain lipid/protein ratios and phospholipid subtype ratios which differ from one another. The identification of sialic acid residues, believed to be limited to the sarcolemma, may also be of value in this approach. We feel that such studies are critical to the characterization of subcellular membranes and, if completed, would add significantly to our understanding of potential calcium-binding sites.

As noted above, the use of density gradient procedures presents difficulties in vascular smooth muscle when comparing gradient data. The data of Wei *et al.* (1976b), Wuytack *et al.* (1978), and Thorens and Haeusler (1978), involving rat mesenteric artery, pig coronary artery, and rabbit aorta, respectively, serve as an interesting example. The first two groups have suggested that cell membrane material appears predominantly in the lighter sucrose bands, i.e., between 28 and 33%, and sarcoplasmic reticulum material in the heavier regions, between 35 and 40%. The latter study suggested that sarcolemma actually appears in heavier fractions than the sarcoplasmic reticulum. All the groups used

5′-nucleotidase as a cell membrane marker, whereas Wei *et al.* (1976b) were the only ones to assay Na^+,K^+-ATPase and found an enrichment in their light fraction. Our work (Table II) further confounds this problem with the measurement of Na^+,K^+-ATPase activity in both the heavy (40%) and light (30%) fractions. Either 5′-nucleotidase is not an adequate membrane marker for vascular smooth muscle, or the transport enzyme is present in areas of the cell other than the cell membrane, or homogenization methods or vessel and species differences are critical. The fact that it is not known if 5′-nucleotidase is an ecto- or endoenzyme in this tissue (Woo and Manery, 1975), and the knowledge that vesicular fractions present substrate accessibility problems pertaining to sidedness may also render questionable the use of 5′-nucleotidase as a marker.

Recent data from our laboratory (Bukoski *et al.*, 1981) have further addressed this problem. We compared distribution of [^{3}H]ouabain binding sites with 5′-nucleotidase activity in subcellular fractions of canine renal and femoral arteries. It was found that the two activities did not coincide in their distribution, suggesting that they are separable. Our assumption was that [^{3}H]ouabain binds only to Na^+ pump sites.

We have been successful in isolating a membrane fraction from canine aorta that has a ouabain-inhibited Na^+,K^+-ATPase activity of 8 μmol P_i/mg protein per hour along with a Mg-ATPase activity of less than 2 μmol P_i/mg protein per hour. This is the most homogeneous Na^+,K^+-ATPase preparation from vascular smooth muscle recorded to date. Although we have succeeded in demonstrating Na^+,K^+-ATPase activity, it has necessitated detergent treatment. Subsequent electron microscopic examination of this fraction indicated it to be nonvesicular in nature, limiting the study of calcium-binding sites and uptake.

Our laboratory has also taken a slightly different approach to the study of calcium binding to membrane fractions. We do not presume that either sarcolemma or sarcoplasmic reticulum can be of more importance in calcium sequestration. We used a fraction known to be heterogeneous, obtained by centrifugation of an initial homogenate at 12,000 *g*, followed by centrifugation of the resulting 12,000-*g* supernatant at 100,000 *g*. This pellet probably contained at least mitochondria, sarcoplasmic reticulum, and sarcolemma. The fraction demonstrated very little Na^+,K^+-ATPase activity identifiable by ouabain inhibition. It demonstrated considerable calcium uptake activity, both ATP-dependent and oxalate-stimulated. This heterogeneous fraction also contained a significant amount of [^{3}H]ouabain binding, suggesting that, despite the fact that no Na^+,K^+-ATPase was demonstrable, sodium pump sites were present. Subsequent treatment of this fraction with SDS manifested significant Na^+,K^+-ATPase activity.

Previous studies have suggested that the lack of demonstrable Na^+,K^+-ATPase activity indicates that Na^+,K^+-ATPase is not present (Verity and Bevan, 1969; Hurwitz *et al.,* 1973; Thorens and Haeusler, 1978; all with rabbit aorta). The fact that detergent treatment can reveal Na^+,K^+-ATPase activity suggests that their conclusions are inaccurate. The question then arises as to how to study calcium binding and uptake in an SDS-treated fraction. The SDS induces permeability changes in the vesicles, which in turn alter oxalate-sensitive calcium uptake. Some preliminary data found in our laboratory may be of significance in understanding the mechanism of regulation of vascular smooth muscle contractility.

In the very heterogeneous preparations, in attempting to identify the coexistence of ouabain binding and Na^+,K^+-ATPase along with ATP-dependent calcium binding and uptake, we have shown a significant amount of ligand-independent calcium binding, i.e., ^{45}Ca binding that requires no additional material except buffer. This binding is inhibited by magnesium, as well as sodium and potassium, and may well mirror the previously published data of Weiner *et al.* (1979), who showed that magnesium could interfere with ^{45}Ca accumulation by intact vessels. We may be identifying specifically a calcium-binding site in vascular smooth muscle preparations that does not require ATP or magnesium but may be released by specific agonists. There is even the possibility that the capacity of this calcium-binding material varies from cell type to cell type. In a membrane fraction, one could identify calcium binding that required no magnesium or ATP, and the addition of magnesium to this calcium-binding fraction would inhibit the ^{45}Ca binding via competition. With the addition of ATP, a stimulation of binding would be observed because of energizing an energy-dependent uptake mechanism, and this ATP-dependent calcium uptake could be further stimulated by oxalate. Thus these heterogeneous fractions have at least three types of calcium sequestration: ATP-dependent, ATP-dependent and oxalate-stimulated, and ligand-independent. It is conceivable that the interaction of these three types of calcium binding in variable degrees from tissue to tissue results in vascular smooth muscle heterogeneity and may be one basis for explaining this well-known phenomenon.

In recent years, some attention also has been paid to the concept of sodium-regulated calcium movement in vascular smooth muscle. The status of this potential mechanism is addressed in the recent review by van Breemen *et al.* (1979). There is presently considerable discussion as to whether such mechanisms are operative in this tissue and, as indicated in that review, much work must be done. It should be briefly mentioned that such a sodium–calcium exchange mechanism has been identified in cardiac muscle sarcolemmal fractions (Reeves and Sutko, 1979; Pitts, 1979). Our laboratory has used techniques similar to those used for cardiac muscle and has not been successful in demonstrating such a phenomenon in vascular smooth muscle membrane fractions.

Any tentative conclusion regarding calcium handling by subcellular fractions of vascular smooth muscle must take many factors into consideration. First, mitochondria, sarcoplasmic reticulum, and cell membrane are all capable of sequestering calcium. In all probability, mitochondria are not involved in the direct contraction–relaxation cycle of vascular smooth muscle. At the present time, it is virtually impossible to say which of the membrane areas, sarcoplasmic reticulum or sarcolemma, is the rate-limiting site in vascular smooth muscle calcium regulation. Indeed, there are hypotheses indicated at the beginning of this chapter in which this differential limiting effect selects for a pharmacologically heterogeneous system, i.e., some vascular smooth muscles will have a rate-limiting site contained in the cell membrane, whereas others will have a rate-limiting calcium-sequestration site contained in the sarcoplasmic reticulum. It must be determined where these rate-limiting steps exist before information may be obtained in order to postulate a molecular mechanism for pharmacological heterogeneity.

IV. CYCLIC ADENOSINE 3′,5′-MONOPHOSPHATE AND CALCIUM REGULATION

A number of workers have suggested that cyclic adenosine 3′,5′-monophosphate (cAMP) stimulates calcium uptake into the microsomal vesicles of vascular smooth muscle (Baudouin-Legros and Meyer, 1973; Bhalla *et al.,* 1978a; Webb and Bhalla, 1976a), thereby giving some biochemical and mechanical basis for the relaxation effect of adrenergic agents on vascular smooth muscle (Table III). Here the suggestion is that relaxation occurs through stimulation of adenyl cyclase, causing an increase in cAMP that indirectly increases calcium sequestration in sarcoplasmic reticulum. The increased sequestration presumably results from a protein kinase-mediated phosphorylation associated with the sarcoplasmic reticulum.

At the present time, the quantitative aspects suggest that cAMP involvement is somewhat unlikely. There are large variations in cAMP content in vascular smooth muscle from 0.16 to 13.3 nmol/g tissue. Whether such variation depends on experimental techniques or actually reflects tissue differences is unclear at the present time (Namm and Leader, 1976). In addition, only relatively large concentrations of cAMP

TABLE III

Effect of cAMP on Calcium Sequestration

Vessel	Fraction[a]	Method[b]	Ca Sequestration: ATP	Ca Sequestration: Oxalate	Ca Sequestration: [Ca]	Ca^{2+}-ATPase	Marker[c]	cAMP[d]	Reference
Rabbit aorta	M	1	5 nmol/mg · 10 min	—	100 μ*M*	5 μmol P/mg · 10 min	Cyt *c* ox.	10^{-4} *M* db-cAMP yields a 20% increase	d'Auriac *et al.*, 1972
Rabbit aorta	M	1	10 nmol/g max	No effect	20 μ*M*	—	Cyt *c* ox.	10^{-5} *M* db-cAMP yields a 25% increase	Baudouin-Legros and Meyer, 1973
Rat aorta	M	1	42 nmol/mg · 10 min	260 nmol/mg · 10 min	10 μ*M*	0.9 μmol P/mg · 10 min	Cyt *c* ox.	16.7 nmol/mg, 5 min to 23.6 nmol/mg, 5 min with 5 μ*M* cAMP	Webb and Bhalla, 1976b
Human umbilical artery	M	1	1 nmol/mg · 10 min	11 nmol/mg · 10 min	100 μ*M*	—	—	10^{-6} *M* cAMP + PK had no effect	Clyman *et al.*, 1976
						—	—	10^{-6} *M* cAMP + PK had no effect	
	Mt	1[c]	17 nmol/mg · 10 min	11 nmol/mg · 10 min	100 μ*M*				
Bovine aorta	LM	1	4 nmol/mg · 15 min	36 nmol/mg · 15 min	0.75 μ*M*	—	—	10^{-6} *M* cAMP ± PK had no effect	Sands *et al.*, 1977
	HM	1	2.4 nmol/mg · 15 min	7.9 nmol/mg · 15 min	0.75 μ*M*	—	—	10^{-6} *M* cAMP ± PK had no effect	
Canine aorta	M	1	60 nmol/mg max	No effect	10–100 μ*M*	15 μmol P/mg	—	10^{-6} *M* cAMP had no effect	Allen, 1977
Rabbit aorta	P	1, 2	7 nmol/mg · 10 min	47 nmol/mg · 10 min	10 μ*M*	—	NKA, 5', G-6-P'tase NADH ox., cyt *c* ox.	10^{-6} *M* cAMP had no effect, but + PK yields a 30% increase	Thorens and Haeusler, 1978
	E	1, 2	15 nmol/mg · 10 min	27 nmol/mg · 10 min	10 μ*M*	—		10^{-6} *M* cAMP had no effect, but + PK yields a 78% increase	
Rat aorta	M	1	—	3.5 nmol/mg · 10 min	10 μ*M*	—	Cyt c ox., SDH	3.5 nmol/mg, 10 min, to 5.3 nmol/mg, 10 min, with 10^{-6} *M* cAMP	Bhalla *et al.*, 1978b

[a] M, Microsomes; Mt, mitochondria, LM, light microsomes; HM, heavy microsomes; P, plasma membrane; E, endoplasmic reticulum.
[b] 1, Differential centrifugation; 2, sucrose density gradient.
[c] Cyt *c* ox., Cytochrome *c* oxidase; NKA, Na^+,K^+-ATPase; G-6-P'tase, glucose-6-phosphatase; SDH, succinic dehydrogenase.
[d] db-cAMP, Dibutyryl cAMP; PK, protein kinase.

(10^{-5}–10^{-6} *M*) stimulate calcium binding to or uptake by vascular smooth muscle microsomal preparations (Baudouin-Legros and Meyer, 1973), and about 20 nmol calcium/mg protein has been the highest stimulation achieved with vascular smooth muscle microsomes. Total tissue cAMP has been known to increase two to three times after the administration of large amounts of beta-adrenergic agents (Namm and Leader, 1976). Consequently, these quantitative aspects make it difficult to assign a cAMP-stimulated calcium uptake role to sarcoplasmic reticulum. In addition, our laboratory has been unable to detect any increase in the calcium sequestration of aortic microsomes, despite a significant protein kinase-dependent, cAMP-stimulated phosphorylation (Allen, 1977). Sands *et al.* (1977) found no phosphorylation of membrane proteins or an effect of cAMP and protein kinase on calcium uptake by bovine or rabbit aortic microsomes (Table III).

A more recent report by Thorens and Haeusler (1978) added more complexity to an already confusing situation. These workers actually found an inhibition of calcium uptake (with oxalate present) in rabbit aorta gradient-isolated microsomes by both cAMP and cGMP (cyclic guanosine 3′-5′-monophosphate). The addition of protein kinase then stimulated uptake by the presumed sarcolemmal fraction more than the sarcoplasmic reticulum fraction. They also found that cyclic nucleotide-stimulated phosphorylation was increased more than calcium uptake, but there was no clear correlation in either fraction between these effects and stimulation of calcium uptake. They concluded by suggesting that, in vascular smooth muscle, a soluble rather than a particulate protein kinase is involved in the regulation of intracellular calcium concentration. Obviously, much work is needed to clarify the relationship among membrane fractions, cAMP, protein kinase, and calcium regulation.

This abbreviated discussion of cAMP effects indicates that perhaps the assignment of sarcoplasmic reticulum as a central factor in a cAMP-mediated calcium sequestration mechanism is partially inconsistent with the currently available data and therefore is premature. In addition, such speculative conclusions may discourage the study of other cellular areas where cAMP may have an important modulating role. For example, Somlyo *et al.* (1972) (for vascular smooth muscle) and Scheid *et al.* (1979) (for stomach cells) have suggested that cAMP effects relaxation by hyperpolarization due to the stimulation of Na^+,K^+-ATPase. Thus the assumption that cAMP acts to stimulate calcium sequestration of microsomes may lead to potentially erroneous conclusions.

Investigators have begun to question even the specific alterations in cyclic nucleotides with smooth muscle contractile states, and relationships once taken for granted are now being examined more carefully.

For example, Diamond and co-workers (Diamond and Hartle, 1976; Diamond and Blisard, 1976) studied the relationship between contractile state and cyclic-nucleotide levels in drug-induced contractions of rat myometrium and drug-induced contractions of canine femoral artery. Contractions of rat myometrium induced by KCl were accompanied by increases in cAMP and decreases in cGMP levels. Both KCl-contracted myometrial strips and phenylephrine-contracted arterial strips could be relaxed by papaverine and nitroglycerine which did not change cAMP levels but increased cGMP levels. Low doses of isoproterenol markedly relaxed the depolarized myometrial preparation, while only causing very small changes in cAMP.

Cohen *et al.* (1977) studied the relationship among age, relaxation, cAMP content, and adenyl cyclase activity of rat aorta. They found that isoproterenol increased cAMP levels to the same extent in all rats at times when the drug was less effective in maximally relaxing aortic strips from older rats. They concluded that decreased relaxation with age was not associated with a reduced ability of vascular relaxants to increase cAMP levels. We believe that there has been no definitive demonstration of a relationship between cAMP and the relaxation of vascular smooth muscle.

The recent review by Kramer and Hardman (1980) arrives at a similar conclusion: "The available evidence, although indicating a general association between cAMP levels and vascular relaxation, does not establish that cAMP is necessary to mediate agonist-induced vascular relaxation. . . ." The possible involvement of cAMP and cAMP-dependent protein kinase in regulating calcium uptake by subcellular fractions from blood vessels is still unclear, and the evidence pertaining to this issue is still conflicting.

V. HYPERTENSION AND SUBCELLULAR CALCIUM REGULATION

A natural extension of recent advancements in the biochemistry of membrane fractions of vascular smooth muscle has been the study of their possible involvement in the development or maintenance of hypertension. A number of laboratories have suggested that the calcium-sequestering capabilities of isolated microsomal fractions of vascular smooth muscle of hypertensive animals are impaired when these fractions are compared to those isolated from vascular smooth muscle of control animals (Table IV). The fundamental suggestion has been that, because of this defect in calcium sequestration, increased calcium

TABLE IV

Membrane Calcium Sequestration and Hypertension

Vessel	Fraction[a]	Method[b]	Ca Sequestration[c]			Ca^{2+} ATPase	Markers[d]	Reference
			ATP	Oxalate	[Ca]			
Rat aorta	M	1	NR: 39.4 nmol max.	—	160 μM	2.22 nmol P/mg · 1 min	—	Aoki *et al.*, 1974
			SHR: 23.6 nmol max.	—	160 μM	2.94 nmol P/mg · 1 min	—	
Rat	M	1	NR: —	24 nmol/mg · 10 min	20 μM	—	5′, NADH ox., cyt *c* ox.	Moore *et al.*, 1975
			NWR: 8 nmol/mg · 10 min	18 nmol/mg · 10 min	20 μM	—		
			SHR: —	14 nmol/mg · 10 min	20 μM	—		
Rat aorta	PM	1, 2	NR: 16.2 nmol/mg · 10 min	No stimulation	100 μM	No activity	PDE I, 5′, cyt *c* ox., K-P'tase	Wei *et al.*, 1976c
			SHR: 11.3 nmol/mg · 10 min	No stimulation	100 μM	No activity		
	SR	1, 2	NR: 14.4 nmol/mg · 10 min	No stimulation	100 μM	No activity		
			SHR: 11.8 nmol/mg · 10 min	No stimulation	100 μM	No activity		
Rat mesenteric artery	PM	1, 2	NR: 15.3 nmol/mg · 10 min	16.0 nmol/mg · 10 min	100 μM	No activity	PDE I, 5′, leucine amino peptidase	Wei *et al.*, 1976b
			SHR: 22.2 nmol/mg · 10 min	23.7 nmol/mg · 10 min	100 μM	—		
	SR	1, 2	NR: 7.72 nmol/mg · 10 min	13.9 nmol/mg · 10 min	100 μM	—		
			SHR: 7.91 nmol/mg · 10 min	15.0 nmol/mg · 10 min	100 μM	—		

(*continued*)

TABLE IV (*Continued*)

Vessel	Fraction[a]	Method[b]	Ca Sequestration[c]			Ca^{2+} ATPase	Markers[d]	Reference
			ATP	Oxalate	[Ca]			
Rat aorta	M	1	—	3.86 nmol/mg · 15 min	10 μM	4.2 nmol P/mg · 30 min		Webb and Bhalla, 1976b
			—	4.36 nmol/mg · 15 min	10 μM	4.2 nmol P/mg · 30 min		
			—	2.11 nmol/mg · 15 min	10 μM	6.5 nmol P/mg · 30 min		
	Mt	1	—	2.98 nmol/mg · 15 min	10 μM	—		
			—	3.23 nmol/mg · 15 min	10 μM	—		
			—	2.99 nmol/mg · 15 min	10 μM	—		
Rat mesenteric artery	PM	1, 2	NWR: 17.2 nmol/mg · 10 min	—	100 μM		Alk. P'tase	Wei *et al.*, 1977b
			SHR: 22.4 nmol/mg · 10 min	—	100 μM		Mg^{2+}-ATPase	
	SR	1, 2	NWR: 11.8 nmol/mg · 10 min		100 μM		Alk. P'tase	
			SHR: 11.7 nmol/mg · 10 min		100 μM		Mg^{2+}-ATPase	
Rat aorta	M	1	NWR: —	3.2 nmol/mg · 10 min	10 μM	—	Cyt *c* ox., SDH	Bhalla *et al.*, 1978b
			SHR: —	1.7 nmol/mg · 10 min	10 μM	—	Cyt *c* ox., SDH	
Rat aorta	M	1	WKY: —	58 nmol/mg · 30 min	20 μM	—	—	Mushlin *et al.*, 1978
			SHR: —	35 nmol/mg · 30 min	20 μM	—	—	
Rat mesenteric artery	PM	1, 2	NWR: 16 nmol/mg · 10 min	—	100 μM	—	Alk. P'tase	Kwan *et al.*, 1980
			KWR: 15 nmol/mg · 10 min	—	100 μM	—	Alk. P'tase	
			SHR: 10 nmol/mg · 10 min	—	100 μM	—	Alk. P'tase	

[a] M, Microsomes; PM, plasma membrane; SR, sarcoplasmic reticulum; Mt, mitochondria.

[b] 1, Differential centrifugation; 2, sucrose density gradient.

[c] NR, Normotensive rats; SHR, spontaneously hypertensive rats; NWR, normotensive Wistar rats; WKY, Wistar–Kyoto rats.

[d] 5′, 5′-Nucleotidase; NADH ox., NADH oxidase; cyt *c* ox., cytochrome *c* oxidase; PDE I, phosphodiesterase; K-P'tase; K phosphatase; alk. P'tase, alkaline phosphatase; SDH, succinic dehydrogenase.

availability to other cellular structures, e.g., contractile proteins, may result in the observed hyperreactivity of smooth muscle from hypertensive animals. This is consistent with suggested alterations of calcium movements that occur in arteries of hypertensive rats. In most cases studied thus far, membrane preparations from hypertensive animals have demonstrated a decreased ability to sequester calcium in response to ATP (Aoki *et al.*, 1974; Moore *et al.*, 1975; Wei *et al.*, 1976c; Webb and Bhalla, 1978a; Mushlin *et al.*, 1978; Kwan *et al.*, 1980; Table IV). At present, however, we feel that any correlations between hypertension and the calcium-sequestering capabilities of vascular smooth muscle are premature and should be viewed with caution. There are a number of reasons for this conservative approach. The first and most important reason is that interpretation of the significance of the data listed in Table IV is extremely difficult, for reasons of fraction heterogeneity and the use of differing arteries as well as variation in control animals used, since the proper control for SHR rats is disputed (McGiff and Quilley, 1981).

The problem involving fraction heterogeneity discussed in detail in the previous section also maintains for the discussion of hypertensive preparations. Workers have not positively identified the presence or absence of different types of membranes in their fractions. Much of the work done on microsomes has suggested that cell membrane material is not present, and therefore any identifiable alteration would occur in sarcoplasmic reticulum. This has not been definitively shown. Furthermore, Daniel and his co-workers have completely excluded the use and importance of sarcoplasmic reticulum in the regulation of contractility in vascular smooth muscle, based largely on the findings of Devine *et al.* (1973) that selected arteries and veins have very small quantities of sarcoplasmic reticulum and on their own data suggesting a low calcium-sequestering level of sarcoplasmic reticulum fragments and little alteration by hypertension. Alternative explanations could be that a large quantity of calcium-sequestering sites are destroyed by their "enrichment" procedure. These workers have yet to perform recovery studies on vascular smooth muscle, and their values for calcium sequestration are quite low when compared to those for other preparations. Daniel and his colleagues have presumed that the cell membrane area is the major regulatory membrane for calcium in vascular smooth muscle. Consequently, any of the later studies they have reported have used a cell membrane-enriched fraction. A recent article by this group (Kwan *et al.*, 1980) retracted earlier suggestions that rat mesenteric arteries show different subcellular alterations in hypertension than aorta from the same hypertensive animal; compare Wei *et al.* (1976b) with Kwan *et al.* (1980) (Table IV). They now believe that their previously presented data, i.e., an increased calcium uptake capacity of the plasma membrane fraction of hypertensive mesenteric arteries, was due to contamination by adhering mesenteric and fat tissue. When the contaminating tissue is adequately removed from the preparation, their data are consistent with all the other data; that is, they show a decrease in calcium uptake in the hypertensive state. It should also be pointed out that other workers have presumed the calcium-sequestering deficiency to reside in sarcoplasmic reticulum rather than sarcolemma, based on their microsomal data (Table IV). However, all these data could be explained by the simple appearance of more inactive protein in a given fraction. Since calcium binding and uptake is measured as a protein concentration-dependent parameter, the simple introduction of inactive protein would decrease calcium binding and uptake significantly. This does not, however, suggest that there may not be alterations in the cell membrane function or sarcoplasmic reticulum function in hypertension. Furthermore, it is not clear whether the data obtained by various workers regarding altered calcium sequestration capabilities are in fact a cause of hypertension or a result of alterations in the vasculature caused by other factors.

It should be emphasized that, at this point, there is still considerable controversy with regard to the etiology of hypertension and whether development of the disease involves altered physical factors of the resistance vessels resulting in changes in the lumen diameter, or whether the increased pressure may partially be due to altered biochemical characteristics regulating excitation–contraction and relaxation. Some workers have shown that veins from hypertensive animals also have altered contractile characteristics similar to those which occur in isolated arteries (Greenberg and Bohr, 1975). Indeed, this suggests that general biochemical alterations in the vasculature may be involved, since the veins are not subjected to the increased pressure. In some physically induced models of hypertension, protected arteries that are not affected by increased pressure still develop biochemical or pharmacological changes (Hansen and Bohr, 1975). A recent paper by Altman *et al.* (1977) provides further data which may indicate that hypertension is a specific vascular manifestation of a more general smooth muscle alteration. These workers tested the contractile response of visceral smooth muscle from spontaneously hypertensive rats (SHRs) using appropriate normotensive Wistar control rats and concluded that SHR fundal smooth muscles showed the same modification in reactivity to barium, strontium, calcium, and diazoxide that was previously described by Janis and Triggle (1973) for arterial smooth muscle from SHRs. These authors feel that these data indicate that cellular modification responsible for the increase in vascular tone in SHRs is not an adaptation reaction to high blood pressure but rather a smooth muscle biochemical alteration of a more general nature.

There are at least four possible alterations in contractility of vascular smooth muscle from hypertensive animals. The first is the increased sensitivity of some vascular smooth muscle to administration of norepinephrine and potassium chloride. This would be difficult to reconcile with simply a generalized increase in available activator calcium to contractile proteins, since the increased sensitivity of vascular smooth muscle is specific to these two agents and other agonists such as angiotensin do not demonstrate such effects (Khosla *et al.,* 1979).

The second altered mechanical characteristic of vascular smooth muscle is decreased maximum contractile output. Again, it would be very difficult to reconcile a general increase in calcium availability as being consistent with a decreased maximal output without postulating an additional inhibitory effect of calcium on other muscle functions.

The third alteration is an increase in spontaneous contractions in arteries from hypertensive animals. Since calcium is known to stabilize the cell membrane, such alterations in the basal electrophysiological characteristics of the muscle cell may be due to alterations in calcium metabolism. Hermsmeyer (1976) has suggested significant increases in the sodium pump characteristics of hypertensive vessels from SHRs. Possible pump involvement has been reinforced by the suggestion of Haddy *et al.* (1978) and Overbeck *et al.* (1976) that ouabain-sensitive ^{86}Rb uptake is significantly depressed in blood vessels from animals with low renin hypertension. This has been expanded recently to suggest that there may be a factor circulating in the blood of these animals, which inhibits active cellular transport. This inhibition then effects increased contractility in the vascular tissue, leading to increased resistance (Haddy *et al.,* 1978). In 1978, Webb and Bohr suggested that potassium-induced vasodilatation in vascular smooth muscle was a measure of sodium pump activity. They later demonstrated that this activity was significantly altered in arteries from hypertensive animals (Webb and Bohr, 1980).

A fourth altered mechanical characteristic of vascular smooth muscle of hypertensive animals which is still largely controversial is a decreased rate of relaxation (Cohen and Berkowitz, 1976). The measurement of relaxation rates is very complex in intact smooth muscle, but this appears to be the only mechanical characteristic consistent with only a decreased calcium uptake by sarcoplasmic reticulum or cell membrane, thereby increasing cell availability of the ion. It is unlikely that increased calcium availability can lead to all the changes noted in hypertensive vessels. A complex disease process such as hypertension (Khosla *et al.,* 1979) probably involves many cellular structures of vascular smooth muscle in much the same manner as heart failure or ischemia involves organelles of cardiac muscle. Furthermore, it is virtually impossible at this time to suggest any similar mechanisms of hypertension development and maintenance among the myriad of experimental models presently in use. Caution must be exercised in interpretation of data suggesting any type of organelle involvement. We feel very strongly that the extrapolation of methods from normotensive to hypertensive animals is premature and that, at present, no definitive statement can be made about subcellular involvement in hypertension.

VI. SUMMARY AND CONCLUSIONS

Recent years have seen significant development in attempting to understand the molecular mechanisms involved in the regulation of cytoplasmic calcium in vascular smooth muscle. Such advances in understanding have been largely due to the analysis of various subcellular fractions from these tissues. Despite the ready identification of calcium-accumulating properties of these fractions in vascular smooth muscle, correlation of the data with whole-tissue ion flux studies and ultrastructural investigations still provides little insight into molecular mechanisms of cellular calcium regulation. Expertise in all three of these areas (subcellular biochemistry, ultrastructure, and ion flux) must be increased in order to understand how vascular smooth muscle is regulated. However significant these advances have been, the specific designation of roles of either sarcolemma or sarcoplasmic reticulum in the regulation of calcium is grossly premature. Problems still exist with isolated fractions, and obviously the potential problems that can arise with overinterpretation of data are enormous. Many workers in this area have been too quick to correlate preliminary data with similar but more consistent data from skeletal or cardiac muscle. However, unlike the situation in striated muscle, much more work is required before a unifying molecular hypothesis of smooth muscle contraction and relaxation can be constructed. Because of these present limitations, three specific areas addressed in this review must be viewed cautiously with respect to molecular mechanism and subcellular organelle involvement: (1) vascular smooth muscle heterogeneity, (2) hypertension, and (3) cAMP-related affects. The latter two areas have unfortunately often suffered from over-interpreted data.

In the cAMP field, publications of about 5 years ago espoused the nucleotide's effect on calcium sequestration as a specific mechanism of beta-adrenergic relaxation, whereas today it is not even clear if its levels correspond to a tissue's contractile state (Kramer and Hardman, 1980). This is not to say that there is no relationship, only that positive identification is premature at this time.

With respect to the involvement of calcium sequestration sites in hypertension, it appears premature to make such positive statements. As indicated in the text, few of the known vascular mechanical alterations can be explained by such a simple scheme.

We have introduced the hypothesis in this review that vascular smooth muscle heterogeneity can be explained by differences in the biochemistry of subcellular membranes involved in calcium regulation, although data are unavailable to support this. Nevertheless, the data that have become available in the past 8 years do at least allow statement of the hypothesis. The correct steps are being taken in trying to elucidate the problems, but unfortunately the answers are not readily forthcoming. As far as subcellular membrane separation is concerned, only somewhat different approaches or new technical advances will provide the means for effecting such separations. We are confident in all three areas listed that the scientists who have brought us this far in the knowledge of the subcellular nature of calcium regulation will continue to lead the way in obtaining the proper information to provide future answers to these important regulatory questions.

ACKNOWLEDGMENTS

This work was supported in part by grants HL 07282 and HL 24585. We would like to acknowledge the helpful comments and critical discussions of Drs. Charles L. Seidel and Francis J. Haddy, and the expert typing of Ms. Loris F. Barrett.

REFERENCES

Adams, H. R., and Goodman, F. R. (1975). Differential inhibitory effect of neomycin on contractile responses of various canine arteries. *J. Pharmacol. Exp. Ther.* **193,** 393–402.

Allen, J. C. (1977). Ca^{2+} binding properties of canine aortic microsomes: Lack of effect of c-AMP. *Blood Vessels* **14,** 91–104.

Allen, J. C., and Seidel, C. L. (1978). EGTA stimulated and ouabain inhibited ATPase of vascular smooth muscle. *In* "Excitation–Contraction Coupling in Smooth Muscle" (R. Casteels, T. Godfraind, and J. C. Rüegg, eds.), pp. 211–218. Elsevier/North Holland Biomedical Press, New York.

Altman, J., DuPonte, D., and Worcel, M. (1977). Evidence for a visceral smooth muscle abnormality in Okamoto spontaneous hypertension. *Br. J. Pharmacol.* **59,** 621–625.

Altura, B. M., and Altura, B. T. (1978). Factors affecting vascular responsiveness. *In* "Microcirculation" (G. Kaley and B. M. Altura, eds.), pp. 547–615. University Park Press, Baltimore, Maryland.

Aoki, K., Ikeda, N., Yamashita, K., Tazumi, K., Sato, I., and Hotta, K. (1974). Cardiovascular contraction in spontaneously hypertensive rat: Ca^{2+} interaction of myofibrils and subcellular membrane of heart and arterial smooth muscle. *Jpn. Circ. J.* **38,** 1115–1121.

Baudouin, M., and Meyer, P. (1972). Calcium release induced by interaction of angiotensin with its receptors in smooth muscle cell microsomes. *Nature (London)* **235,** 336–338.

Baudouin-Legros, M., and Meyer, P. (1973). Effects of angiotensin, catecholamines and cyclic AMP on calcium storage in aortic microsomes. *Br. J. Pharmacol.* **47,** 377–385.

Besch, H. R., Jones, L. R., and Watanabe, A. M. (1976). Intact vesicles of canine cardiac sarcolemma: Evidence from vectorial properties of Na^+, K^+-ATPase. *Circ. Res.* **39,** 586–595.

Bhalla, R. C., Webb, R. C., Singh, D., and Brock, T. (1978a). Role of cyclic AMP in rat aortic microsomal phosphorylation and calcium uptake. *Am. J. Physiol.* **234,** H508–H514.

Bhalla, R. C., Webb, R. C., Singh, D., Ashley, T., and Brock, T. (1978b). Calcium fluxes, calcium binding and adenosine cyclic 3′,5′-monophosphate-dependent protein kinase activity in the aorta of spontaneously hypertensive and Kyoto-Wistar normotensive rats. *Mol. Pharmacol.* **14,** 468–477.

Bohr, D. F., Sitrin, M. D., and Sobieski, J. (1971). Heterogeneity among vascular smooth muscles in the regulation of activator calcium. *In* "Physiology and Pharmacology of Neuroeffector Systems" (J. A. Bevan, R. F. Furchgott, R. A. Maxwell, and A. P. Somlyo, eds.), pp. 72–85. Karger, Basel.

Boyer, P. D., Chance, B., Ernster, L., Mitchell, P., Racker, E., and Slater, E. C. (1977). Oxidative phosphorylation and photophosphorylation. *Annu. Rev. Biochem.* **46,** 955–1026.

Brading, A. F., and Widdicombe, J. (1974). An estimate of sodium/potassium pump activity and the number of pump sites in the smooth muscle of the guinea-pig taenia coli, using [^{3}H]-ouabain. *J. Physiol. (London)* **238,** 235–249.

Bukoski, R. D., Seidel, C. L., and Allen, J. C. (1979). Effect of ionophore RO 2-2985 on the contractile response of canine coronary, renal and femoral arteries. *Blood Vessels* **16,** 281–294.

Bukoski, R. D., Seidel, C. L., and Allen, J. C. (1981). Heterogeneous distribution and activity of subcellular markers in renal and femoral arteries. *Fed. Proc.* **40,** 624. (abstr).

Clyman, R. I., Manganiello, V. C., and Lovell-Smith, C. J. (1976). Calcium uptake by subcellular fractions of human umbilical artery. *Am. J. Physiol.* **231,** 1074–1081.

Cohen, M. L., and Berkowitz, B. A. (1976). Decreased vascular relaxation in hypertension. *J. Pharmacol. Exp. Ther.* **196,** 396–406.

Cohen, M. L., Blume, A. S., and Berkowitz, B. A. (1977). Vascular adenylate cyclase: Role of age and guanine nucleotide activation. *Blood Vessels* **14,** 25–42.

Crompton, M. V., Capano, M., and Carafoli, E. (1976). Respiration-dependent efflux of magnesium ions from heart mitochondria. *Biochem. J.* **154,** 735–742.

Crompton, M. V., Moser, R., Ludi, H., and Carafoli, E. (1978). The interrelations between the transport of sodium and calcium in mitochondria of various mamalian tissues. *Eur. J. Biochem.* **82,** 25–31.

d'Auriac, G., Baudouin, M., and Meyer, P. (1972). Mechanism of action of angiotensin in smooth muscle cell. *Circ. Res.* **30–31,** Suppl. II, II-151–II-160.

Devine, C. E., Somlyo, A. V., and Samlyo, A. P. (1973). Sarcoplasmic reticulum and mitochondria as cation accumulation sites in smooth muscle. *Philos. Trans. R. Soc. London, Ser. B* **265,** 17–23.

Devynck, M.-A., Pernollet, M. G., Meyer, P., Fermandjian, S., and Fromagot, P. (1973).

Angiotensin receptors in smooth muscle cell membranes. *Nature (London) New Biol.* **245,** 55–58.

Diamond, J., and Blisard, K. S. (1976). Effects of stimulant and relaxant drugs on tension and cyclic nucleotide levels in canine femoral artery. *Mol. Pharmacol.* **12,** 688–692.

Diamond, J., and Hartle, D. K. (1976). Cyclic nucleotide levels during carbachol-induced smooth muscle contractions. *J. Cyclic Nucleotide Res.* **2,** 179–188.

Fitzpatrick, D. F., Landon, E. J., Debbas, G., and Hurwitz, L. (1972). A calcium pump in vascular smooth muscle. *Science* **176,** 305–306.

Fleming, W. W. (1980). The electrogenic Na^+,K^+-pump in smooth muscle: Physiologic and pharmacologic significance. *Annu. Rev. Pharmacol.* **20,** 129–149.

Forbes, M. S., Rennels, M. L., and Nelson, E. (1979). Calveolar systems and sarcoplasmic reticulum in coronary smooth muscle cells of the mouse. *J. Ultrastruct. Res.* **67,** 325–339.

Ford, G. D., and Hess, M. L. (1975). Calcium-accumulating properties of subcellular fractions of bovine vascular smooth muscle. *Circ. Res.* **37,** 580–587.

Gerthoffer, W. T., and Allen, J. C. (1981). Characteristics of binding of ^3H-ouabain to smooth muscle Na^+,K^+-ATPase and quantitation of Na^+-K^+ pump sites. *J. Pharmacol. Exp. Ther.* **217,** 692–696.

Golenhofen, K. (1976). Spontaneous activity and functional classification of mammalian smooth muscle. *In* "Physiology of Smooth Muscle" (E. Bulbring and M. R. Shuba, eds.), pp. 91–98. Raven Press, New York.

Greenberg, S., and Bohr, D. F. (1975). Venous smooth muscle in hypertension. *Circ. Res.* **36,** Supp. II, 208–215.

Haddy, F., Pamnani, M., and Clough, D. (1978). Review: The sodium-potassium pump in volume expanded hypertension. *Clin. Exp. Hypertens.* **1,** 295–336.

Hanley, H. G., Lewis, R. M., Hartley, C. J., Franklin, D., and Schwartz, A. (1975). Effects of an inotropic agent, RO 2-2985 (X537A), on regional blood flow and myocardial function in chronically instrumented conscious dogs and anesthetized dogs. *Circ. Res.* **37,** 215–225.

Hansen, T. R., and Bohr, D. F. (1975). Hypertension: Transmural pressure and vascular smooth muscle response in rats. *Circ. Res.* **36,** 590–598.

Hansen, T. R., Abrams, G. D., and Bohr, D. F. (1974). Role of pressure in structural and functional changes in arteries of hypertensive rats. *Circ. Res.* **34,** Suppl. I, I-101–I-108.

Hermsmeyer, K. (1976). Electrogenesis of increased norepinephrine sensitivity of arterial vascular muscle hypertension. *Circ. Res.* **38,** Suppl. II, 362–367.

Hess, M. L., and Ford, G. D. (1974). Calcium accumulation by subcellular fractions from vascular smooth muscle. *J. Mol. Cell. Cardiol.* **6,** 275–282.

Hurwitz, L., Fitzpatrick, D. F., Debbas, G., and Landon, E. J. (1973). Localization of calcium pump activity in smooth muscle. *Science* **179,** 384–386.

Janis, R. A., and Triggle, D. R. (1973). Effect of diazoxide on aortic reactivity to calcium in spontaneously hypertensive rats. *Can. J. Physiol. Pharmacol.* **51,** 621–626.

Jones, A. W. (1980). Content and fluxes of electrolytes. *In* "Handbook of Physiology" (D. F. Bohr, A. V. Somlyo, and H. V. Sparks, eds.), Sect. 2, Vol. II, pp. 253–299. Am. Physiol. Soc., Washington, D.C.

Jones, L. R., Besch, H. R., Fleming, J. W., McConnaughey, M. M., and Watanabe, A. M. (1979). Separation of vesicles of cardiac sarcolemma from vesicles of cardiac sarcoplasmic reticulum: Comparative biochemical analysis of component activities. *J. Biol. Chem.* **254,** 530–539.

Khosla, M. C., Page, I. H., and Bumpus, F. M. (1979). Interrelations between various blood pressure regulatory systems and the mosaic theory of hypertension. *Biochem. Pharmacol.* **28,** 2867-2882.

Kramer, G. L., and Hardman, J. G. (1980). Cyclic nucleotides and blood vessel contraction. *In* "Handbook of Physiology" (D. F. Bohr, A. V. Somlyo and H. Sparks, eds.), Sect. 2, Vol. II, pp. 161-178. Am. Physiol. Soc., Washington, D.C.

Kuschinsky, K., Lullmann, H., and Van Zwieten, P. A. (1968). A comparison of the accumulation and release of ^{3}H-ouabain and ^{3}H-digitoxin by guinea-pig heart muscle. *Br. J. Pharmacol. Chemother.* **32,** 598-608.

Kutsky, P., Weiss, G. B., and Karaki, H. (1980). Delineation of high and low affinity ^{45}Ca incorporation and of ^{45}Ca efflux in canine aortic smooth muscle microsomes. Gen. Pharmac. **11,** 475-481.

Kwan, C.-Y., and Daniel, E. E. (1980). Vascular smooth muscle changes in hypertension. *Blood Vessels* **17,** 3.

Kwan, C.-Y., Garfield, R., and Daniel, E. E. (1979a). An improved procedure for the isolation of plasma membranes from rat mesenteric arteries. *J. Mol. Cell. Cardiol.* **11,** 639-660.

Kwan, C.-Y., Belbeck, L., and Daniel, E. E. (1979b). Abnormal biochemistry of vascular smooth muscle plasma membrane as an important factor in the initiation and maintenance of hypertension in rats. *Blood Vessels* **16,** 259-268.

Kwan, C.-Y., Belbeck, L., and Daniel, E. E. (1980). Abnormal biochemistry of vascular smooth muscle plasma membrane isolated from hypertensive rats. *Mol. Pharmacol.* **17,** 137-140.

Lehninger, A. L., Reynafarje, B., Vercesi, A., and Tew, W. P. (1978). Transport and accumulation of calcium in mitochondria. *Ann. N. Y. Acad. Sci.* **307,** 160-176.

McGiff, J. C., and Quilley, C. P. (1981). The rat with spontaneous genetic hypertension is not a suitable model of human essential hypertension. Circ. Res. **48,** 455-463.

MacNamara, D. B., Roulet, M. J., Hyman, A. L., and Kadowitz, P. J. (1979). Characterization of the energy-dependent calcium binding of a mitochondrial fraction isolated from bovine intrapulmonary vein. *Can. J. Physiol. Pharmacol.* **57,** 1107-1113.

Michael, L. H., Schwartz, A., and Wallick, E. T. (1979). Nature of the transport adenosine triphosphatase-digitalis complex. XIV. Inotropy and cardiac glycoside interaction with Na^{+},K^{+}-ATPase of isolated cat papillary muscles. *Mol. Pharmacol.* **16,** 135-146.

Moore, L., Hurwitz, L., Davenport, G. R., and Landon, E. J. (1975). Energy-dependent calcium uptake activity of microsomes from the aorta of normal and hypertensive rats. *Biochim. Biophys. Acta* **413,** 432-443.

Murphy, R. A. (1979). Filament organization and contractile function in vertebrate smooth muscle. *Annu. Rev. Physiol.* **41,** 737-748.

Mushlin, P. S., Sastry, B. V. R., Boerth, R. C., Surber, M. J., and Landon, E. J. (1978). Dithiothreitol-induced alterations of blood pressure, vascular reactivity and aortic microsomal calcium uptake in spontaneously hypertensive rats. *J. Pharmacol. Exp. Ther.* **207,** 331-339.

Namm, D. H., and Leader, J. P. (1976). Occurrence and function of cyclic nucleotides in blood vessels. *Blood Vessels* **13,** 24-47.

Overbeck, H. W. Pamnani, M. B., Akera, T., Brody, T. M., and Haddy, F. J. (1976). Depressed function of a ouabain-sensitive sodium-potassium pump in blood vessels from renal hypertensive dogs. *Circ. Res.* **38,** Suppl. II, 48-52.

Pitts, B. J. R. (1979). Stoichiometry of sodium-calcium exchange in cardiac sarcolemmal vesicles: Coupling to sodium pump. *J. Biol. Chem.* **254,** 6232-6235.

Preiss, R., and Banaschak, H. (1979). Na^{+},K^{+}-ATPase in excitation-contraction coupling of vascular smooth muscle from cattle. *Acta Biol. Med. Ger.* **38,** 83-96.

Reeves, J. R., and Sutko, J. L. (1979). Sodium–calcium ion exchange in cardiac membrane vesicles. *Proc. Natl. Acad. Sci. U.S.A.* **76,** 590–594.

Ross, R., and Kariya, B. (1980). Morphogenesis of vascular smooth muscle in atherosclerosis and cell structure. *In* "Handbook of Physiology" (D. F. Bohr, A. P. Somlyo, and H. V. Sparks, eds.), Sect. 2, Vol. II, pp. 61–92. Am. Physiol. Soc., Washington, D.C.

Sands, H., Mascali, J., and Paietta, E. (1977). Determination of calcium transport and phosphoprotein phosphatase activity in microsomes from respiratory and vascular smooth muscle. *Biochim. Biophys. Acta* **500,** 223–234.

Scarpa, A. (1976). Kinetic and thermodynamic aspects of mitochondrial calcium transport. *In* "Mitochondria: Bioenergetics, Biogenesis and Membrane Structure" (L. Packer and A. Gomez-Puyou, eds.), pp. 31–45. Academic Press, New York.

Scheid, C. R., Honeyman, J. W., and Fay, F. S. (1979). Mechanism of β-adrenergic relaxation of smooth muscle. *Nature (London)* **277,** 32–36.

Schwartz, A., Lindenmeyer, G., and Allen, J. C. (1975). The sodium-potassium adenosinetriphosphatase: Pharmacological, physiological and biochemical aspects. *Pharmacol. Rev.* **27,** 3–134.

Shibata, S., Kurahashi, K., and Kuchii, M. (1973). Possible etiology of contractile impairment of vascular smooth muscle from spontaneously hypertensive rats. *J. Pharmacol. Exp. Ther.* **185,** 406–417.

Sloane, B. F. (1980). Isolated membranes and organelles from vascular smooth muscles. *In* "Handbook of Physiology" (D. F. Bohr, A. V. Somlyo, and H. V. Sparks, eds.), Sect. 2, Vol. II, pp. 121–132. Am. Physiol. Soc., Washington, D.C.

Sloane, B. F., Scarpa, A., and Somlyo, A. (1978). Vascular smooth muscle mitochondria: Magnesium content and transport. *Arch. Biochem. Biophys.* **189,** 409–416.

Somlyo, A. P., and Somlyo, A. V. (1968). Vascular smooth muscle. I. Normal structure, pathology, biochemistry and biophysics. *Pharmacol. Rev.* **20,** 197–272.

Somlyo, A. P., and Somlyo, A. V. (1970). Vascular smooth muscle. II. Pharmacology of normal and hypertensive vessels. *Pharmacol. Rev.* **22,** 249–353.

Somlyo, A. P., Somlyo, A. V., and Smiesko, V. (1972). Cyclic AMP and vascular smooth muscle. *Adv. Cyclic Nucleotide Res.* **1,** 175–194.

Somlyo, A. V. (1980). Ultrastructure of vascular smooth muscle. *In* "Handbook of Physiology" (D. F. Bohr, A. P. Somlyo, and H. V. Sparks, eds.), Sect. 2, Vol. II, pp. 69–91, Am. Physiol. Soc., Washington, D. C.

Thorens, S. (1979). Ca^{2+}-ATPase and Ca uptake without requirement for Mg^{2+} in membrane fractions of vascular smooth muscle. *FEBS Lett.* **98,** 177–180.

Thorens, S., and Haeusler, G. (1978). Effects of adenosine 3′:5′-monophosphate and guanosine 3′:5′-monophosphate on calcium uptake and phosphorylation in membrane fractions of vascular smooth muscle. *Biochim. Biophys. Acta* **512,** 415–428.

Vallieres, J., Scarpa, A., and Somlyo, A. P. (1975). Subcellular fractions of smooth muscle: Isolation, substrate utilization and Ca^{++} transport by main pulmonary artery and mesenteric vein mitochondria. *Arch. Biochem. Biophys.* **170,** 659–669.

Van Breemen, C., Aaronson, P., and Loutzenhiser, R. (1979). Na–Ca interactions in mammalian smooth muscle. *Pharmacol. Rev.* **30,** 167–208.

Vanhoutte, P. M. (1978). Heterogeneity in vascular smooth muscle. *In* "Microcirculation" (G. Kaley and B. M. Altura, eds.), pp. 181–310. University Park Press, Baltimore, Maryland.

Verity, M. A., and Bevan, J. A. (1969). Membrane adenosine triphosphatase activity of vascular smooth muscle. *Biochem. Pharmacol.* **18,** 327–338.

Webb, R. C., and Bhalla, R. C. (1976a). Calcium sequestration by subcellular fractions isolated from vascular smooth muscle: Effect of cyclic nucleotides and prostaglandins. *J. Mol. Cell. Cardiol.* **8,** 145–157.

Webb, R. C., and Bhalla, R. C. (1976b). Altered calcium sequestration by subcellular fractions of vascular smooth muscle from spontaneously hypertensive rats. *J. Mol. Cell. Cardiol.* **8,** 651–661.

Webb, R. C., and Bohr, D. F. (1978). Potassium induced relaxation as an indicator of Na^+,K^+-ATPase activity in vascular smooth muscle. *Blood Vessels* **15,** 198–207.

Webb, R. C., and Bohr, D. F. (1980). Vascular reactivity in hypertension: Altered effect of ouabain. *Experientia* **36,** 220–222.

Wei, J.-W., Janis, R. A., and Daniel, E. E. (1976a). Isolation and characterization of plasma membrane from rat mesenteric arteries. *Blood Vessels* **13,** 279–292.

Wei, J.-W., Janis, R. A., and Daniel, E. E. (1976b). Studies on subcellular fractions from mesenteric arteries of spontaneously hypertensive rats: Alterations in both calcium uptake and enzyme activities. *Blood Vessels* **13,** 293–308.

Wei, J.-W., Janis, R. A., and Daniel, E. E. (1976c). Calcium accumulation and enzymatic activities of subcellular fractions from aortas and ventricles of genetically hypertensive rats. *Circ. Res.* **39,** 133–140.

Wei, J.-W., Janis, R. A., and Daniel, E. E. (1977a). Alterations in calcium transport and binding by the plasma membrane of mesenteric arteries from spontaneously hypertensive rats. *Blood Vessels* **14,** 55–64.

Wei, J.-W., Janis, R. A., and Daniel, E. E. (1977b). Relationship between blood pressure of spontaneously hypertensive rats and alterations in membrane properties of mesenteric arteries. *Circ. Res.* **40,** 299–305.

Wiener, R., Turlapaty P., and Altura, B. M. (1979). Influence of extracellular Mg on contractility and Ca uptake in venous smooth muscle. *Fed. Proc., Fed. Am. Soc. Exp. Biol.* **38,** 1136 (abstr.).

Weiss, G. B. (1977). Calcium and contractility in vascular smooth smooth muscle. *Adv. Gen. Cell. Pharmacol.* **2,** 71–155.

Wolowyk, M. W., Kidwai, A. M., and Daniel, E. E. (1971). Sodium-potassium stimulated adenosinetriphosphatase of vascular smooth muscle. *Can. J. Biochem.* **49,** 376–384.

Woo, Y.-T., and Manery, J. F. (1975). 5′-Nucleotidase: An ecto-enzyme of frog skeletal muscle. *Biochim. Biophys. Acta* **397,** 144–152.

Wuytack, F., and Casteels, R. (1980). Demonstration of a (Ca^{2+} + Mg^{2+})-ATPase activity probably related to Ca^{2+} transport in the microsomal fraction of porcine coronary artery smooth muscle. *Biochim. Biophys. Acta* **595,** 257–263.

Wuytack, F., Landon, E., Fleischer, S., and Hardman, J. G. (1978). The calcium accumulation in a microsomal fraction from porcine coronary artery smooth muscle. *Biochim. Biophys. Acta* **540,** 253–269.

Zelck, U., Karnstedt, U., and Albrecht, E. (1975). Calcium uptake and calcium release by subcellular fractions of smooth muscle. *Acta Biol. Med. Ger.* **34,** 981–986.

5

The Contractile Apparatus of Smooth Muscle and Its Regulation by Calcium

D. J. Hartshorne

I. INTRODUCTION

The intracellular concentration of free Ca^{2+} within a muscle cell regulates the contractile activity of that cell. At approximately 5×10^{-6} *M* Ca^{2+} the muscle contracts, and at lower concentrations, approaching 1×10^{-7} *M*, the muscle relaxes. In skeletal muscle the level of Ca^{2+} available to the contractile apparatus is governed by an intracellular membranous network termed the sarcoplasmic reticulum (SR). The SR releases Ca^{2+} following depolarization of the cell membrane to initiate contraction, and when excitation ceases the SR sequester Ca^{2+} to promote relaxation. It has been shown that SR exists in many types of smooth muscle, and is assumed that its function is similar to that in skeletal muscle; i.e., it acts as

VASCULAR SMOOTH MUSCLE: METABOLIC, IONIC, AND CONTRACTILE MECHANISMS

ISBN 0-12-195220-7

an intracellular Ca^{2+} sink. Whether the smooth muscle SR can supply all the Ca^{2+} necessary for activation of the contractile apparatus, or whether alternate extracellular Ca^{2+} sources are utilized, is a controversial subject and will not be considered here (see Johansson and Somlyo, 1980; Somlyo, 1980; van Breemen *et al.*, 1980; Weiss, 1977). However, it is universally accepted that the level of available Ca^{2+} within all types of muscle cells determines contractile activity, and therefore it follows that the contractile apparatus (i.e., the thick and thin filaments) in these cells must possess a means of sensing intracellular concentrations of Ca^{2+}. This detection mechanism is often referred to as the regulatory system. In this article the types of regulatory mechanisms that have been identified in different muscle types will be discussed briefly, and the emphasis will then shift to a consideration of the regulatory mechanism in vertebrate smooth muscle. Before this is attempted, however, some preliminary remarks are in order concerning the system that is subject to the regulatory process, namely, the contractile apparatus.

II. CONTRACTILE APPARATUS

The conventional view is that the contractile apparatus is composed of thick and thin filaments. This is based, to a large extent, on the situation found with skeletal muscle and dates back to the original observations by A. F. Huxley and Niedergerke (1954) and H. E. Huxley and Hanson (1954) that the shortening of striated muscle was due to the relative sliding between two sets of filaments. Neither the thin (actin-containing) nor the thick (myosin-containing) filaments alter in length, and the length changes of the muscle result from an increased or decreased extent of filament overlap. It has subsequently been realized that tension is generated as a result of the interaction between the myosin cross-bridges (of the thick filament) and actin. In relaxed muscle the cross-bridge contacts are not made, and the two filament types can slide freely past each other. From a biochemical point of view the formation of cross-bridge–actin interactions is correlated with the binding of actin to myosin, and cross-bridge cycling (one cycle per molecule of ATP hydrolyzed) is assumed to be analogous to the activation by actin of the Mg^{2+}-ATPase activity of myosin.

The sliding-filament model has become a fundamental component of any treatment of the contractile process in either skeletal or cardiac muscle, and a similar situation has been sought in smooth muscle. The initial requirement is obviously that both filament types exist. The demonstration of this was not a simple matter in smooth muscle, and a double-filament array was not established until several years after its discovery in striated muscle. Thin filaments are numerous and were observed in many types of smooth muscle, but thick filaments are more labile and were not demonstrated until more refined fixation techniques became available (Devine and Somlyo, 1971; Kelley and Rice, 1968; Nonomura, 1968). However, it is now accepted that both filament types exist in smooth muscle, and the popular dogma is that tension is developed via cross-bridge–actin interactions, as it is in striated muscle. Thus in its simplest concept the mechanism which regulates the activity of the contractile apparatus does so by controlling the interactions of the myosin cross-bridges with actin.

The thin filaments in smooth muscle are abundant and morphologically appear similar to those in skeletal muscle. The length of the thin filaments has not been established for all muscle types, but in chicken gizzard it is suggested that they are shorter than skeletal filaments, being less than 1 μm (Nonomura *et al.*, 1980). The backbone of the filament is a double-helical array of actin molecules, and the only other major protein component of smooth muscle filaments is tropomyosin, present with a stoichiometry of about one tropomyosin to six or seven actins. Troponin is not usually found in smooth muscle thin filaments (Driska and Hartshorne, 1975; Sobieszek and Small, 1976), and this is a significant difference between smooth and striated muscle thin filaments. Recently, however, it has been shown that thin filaments from pig aorta (Marston *et al.*, 1980) contain several components in addition to actin and tropomyosin, some of which appear to be troponin-like. Thus the *in vivo* complement of thin-filament components remains to be established.

The thin filaments seem to be anchored at one end to amorphous structures called dense bodies. The latter are found both in the cytoplasm and attached to the plasma membrane. It is likely, but not proven, that both forms of the dense bodies are analogous to the Z lines of striated muscle (Somlyo, 1980). In support of this, α-actinin, which is known to be a component of Z lines, has been localized in the dense bodies (Schollmeyer *et al.*, 1976). A second filament type is also thought to be associated with the dense bodies, namely, 100Å- or intermediate filaments (Somlyo, 1980; Small and Sobieszek, 1980). The function of these filaments is not established, but it is suggested that they serve as a cytoskeleton. The point to emphasize, however, is that the thin filaments are not "free-floating" and, similar to the thin filaments of skeletal muscle, are locked into some form of intracellular framework. This is of obvious importance for the transmission of tension developed via the interactions between thick and thin filaments.

Thick filaments are now an accepted component of the ultrastructure of the smooth muscle cell and have been described in a variety of smooth muscles (Somlyo, 1980; Small and Sobieszek, 1980). Yet despite numerous observations the detailed morphology of native thick filaments has not been established. For example, the length of the thick filaments is not agreed upon. Ashton *et al.* (1975) reported a length of 2.2 μm for the filaments from rabbit portal anterior mesenteric vein, whereas considerably longer filaments, up to 8 μm, were observed in homogenates of taenia coli (Small, 1977). The way in which myosin molecules are assembled to form the thick filament is also an unresolved but important point. Studies on synthetic filaments, i.e., those formed from myosin *in vitro* (Somlyo, 1980; Small and Sobieszek, 1980), have yielded a variety of filament types. In general, however, these investigations did not identify a specific mode of myosin assembly that was consistent with the observed native filaments. A possible exception to this is the type of assembly described by Sobieszek (1977a) and Hinssen *et al.* (1978). The building unit was thought to be an antiparallel myosin dimer, with an overlap of about 600 Å, which when assembled into an *in vitro* filament bore several similarities to the native filament structure. A final comment on the properties of thick filaments is that they appear to be more stable to dissociation by ATP when formed from phosphorylated myosin (Suzuki *et al.*, 1978). This effect is also observed with nonmuscle myosin (Scholey *et al.*, 1980). Thus one would predict that, in relaxed muscle, where the myosin is largely dephosphorylated, the thick filament structure would be more labile and possibly also subject to the generation of artifactual structures.

A striking feature of the ultrastructure of the smooth muscle cell is the number of thin filaments relative to the number of thick filaments. In rabbit skeletal muscle there are approximately two thin filaments per thick filament, whereas this ratio is considerably higher in smooth muscle. Filament counts of 10:1 to 15:1 have been reported in a variety of smooth muscles (see Hartshorne and Gorecka, 1980). These values are consistent with the high actin content in smooth muscle cells. Analyses carried out with several smooth muscles (Cohen and Murphy, 1978) indicate that roughly two classes of muscle can be distinguished. One is listed as arterial, and the other as nonarterial (consisting of esophagus, trachea, intestine, and uterus muscles). The major difference between the two classes is in the content of actin; arterial muscle contains about 50 mg/g cell wet weight as compared to about 27.5 mg/g for nonarterial muscle. In both classes the myosin content is similar but is considerably less than in skeletal muscle. If one makes a rough estimate of the concentrations of the major contractile proteins, the following values are obtained: for arterial muscle, myosin 0.04 m*M*, actin 1.2 m*M*, tropomyosin 0.2 m*M*; for nonarterial muscle, myosin 0.04 m*M*, actin 0.66 m*M*, tropomyosin, 0.11 m*M*; for skeletal muscle, myosin 0.13 m*M*, actin 0.5 m*M*, tropomyosin 0.07 m*M*. When comparing skeletal to smooth muscle, a general pattern emerges, namely, that the amount of actin relative to myosin is considerably higher in smooth than in skeletal muscle. This is partly because the concentration of myosin is lower in smooth muscle. The latter observation is interesting because it has been shown that the tension developed by smooth muscle is approximately equal to that developed in skeletal muscle (Murphy *et al.*, 1974), and further it is established that tension is proportional to the number of cross-bridges acting in parallel (Huxley, 1957). Thus, when considering tension development in smooth muscle, there must be some way of compensating for the lower content of myosin, and therefore a smaller number of available cross-bridges. Several features have been suggested as a partial explanation, including the longer smooth muscle thick filaments, an altered kinetic cycle in smooth muscle, and the higher content of actin in smooth muscle. These and other possibilities are discussed by Murphy (1979) and Somlyo (1980).

To summarize the forgoing discussion, it is accepted that the contractile apparatus of smooth muscle consists of thick and thin filaments, and it is generally assumed that length changes in smooth muscle are caused by a sliding-filament mechanism. This implies that tension is developed as a result of cross-bridge–actin interactions, and it follows also that the Ca^{2+}-dependent regulatory mechanism functions by controlling these interactions.

III. TYPES OF REGULATORY MECHANISMS

A. Thin-Filament Regulation

In muscle biochemistry most of our knowledge has accumulated for the skeletal muscle system. With respect to the regulatory mechanism this is certainly true, and the control mechanism of skeletal muscle actomyosin is well established. It is known that the regulatory proteins, troponin and tropomyosin, are located on the thin filaments with a molar stoichiometry of one troponin to one tropomyosin to seven actins. Troponin has three different subunits, I, T, and C, and troponin C is the receptor for the activating Ca^{2+}. Troponin T binds to tropomyosin, and troponin I is thought to bind to actin in the absence of Ca^{2+} and thereby stabilize the inhibited, or "off," state. The function of tropomyosin is critical to the regulatory mechanism, and a popular concept (Murray and Weber, 1974) is that tropomyosin is moved from a position which blocks the interaction of the myosin cross-bridge with actin to a nonblocking position. The blocking and nonblocking states correspond to relaxed and contracting muscle, respectively, and the switch for the tropomyosin movement is the binding of Ca^{2+} by troponin C. This model of thin-filament regulation is probably an oversimplication, but it is consistent with many of the experimental observations. It also incorporates the concept of potentiation by tropomyosin, which rationalizes how one tropomyosin molecule by its lateral alignment along the thin filament can influence about seven actin monomers. In order for the movement of tropomyosin, induced at one point in the regulated complex of seven actins, one tropomyosin, and one troponin, to be maintained over about 400 Å (the length of the tropomyosin molecule) the tropomyosin molecule must be fairly rigid. This is consistent with the known structure of tropomyosin in which the two subunits are arranged in a coiled-coil α-helical configuration.

The best known example of thin-filament regulation is found in vertebrate skeletal muscle. Cardiac muscle also is regulated by this type of mechanism. The distinctive feature is that the regulatory components are confined to the thin filament. If actin is prepared without the associated regulatory proteins and assayed with skeletal muscle myosin, then the Mg^{2+}-ATPase activity is not affected by alternations in the Ca^{2+} concentration. Also, the ATPase activity is close to that observed in the presence of Ca^{2+}, troponin, and tropomyosin. Thus the function of the regulatory complex is to inhibit ATPase activity in the absence of Ca^{2+}.

B. Thick-Filament Regulation

Subsequent to the discovery of troponin-linked, or thin-filament, regulation, it was found that in molluscs the regulatory mechanism was confined to the myosin molecule (Kendrick-Jones *et al.*, 1970), and this effect was later shown to be associated with specific myosin light chains (Szent-Györgyi *et al.*, 1973; Kendrick-Jones *et al.*, 1976), a high-affinity Ca^{2+}-binding site being formed by interaction of the myosin heavy chain and the regulatory light chain. This type of regulatory mechanism was also demonstrated in other invertebrates (Lehman and Szent-Györgyi, 1975). Recently it was shown (Chantler *et al.*, 1981) that the activation of ATPase activity of acto-scallop heavy meromyosin (a soluble enzymically active fragment of myosin obtained by limited proteolysis) required the binding of Ca^{2+} to each of the two myosin heads, or active sites, and this indicated an intramolecular cooperativity between the two heads. When whole myosin was used, the activation profile as a function of Ca^{2+} concentration became more complex, and the suggestion was made that this reflected intermolecular cooperativity within the thick filament (Chantler *et al.*, 1981).

Conceptually the myosin-linked system, as outlined above, is the simplest mode of regulation. It can be postulated that the Ca^{2+}-free sites are inhibitory to ATPase activity and that inhibition is relieved on binding Ca^{2+} to these sites. In this context it is pertinent that the desensitization of myosin which occurs on removal of the regulatory light chains results in an active but unregulated state (Szent-Györgyi *et al.*, 1973). In vertebrate skeletal muscle the Ca^{2+}-binding site is shifted to the thin filament, where the Ca^{2+}-free troponin complex is inhibitory. Thus in these two systems regulation is associated either with the thick or the thin filaments [in some instances dual regulation is also found (Lehman and Szent-Györgyi, 1975)]. Szent-Györgyi and colleagues used the difference in the location of the regulatory systems to devise an elegant, yet simple, test for distinguishing between actin- and myosin-linked regulation in various types of muscles (Lehman *et al.*, 1972).

IV. REGULATORY MECHANISMS IN VERTEBRATE SMOOTH MUSCLE

The historical development of this area of biochemistry followed, to a large extent, advances made with other muscle types. Following the discovery of troponin in skeletal muscle (Ebashi, 1963; Ebashi *et al.*, 1968), several studies were reported in which troponin-like components were detected in smooth muscle (Carsten, 1971; Ebashi *et al.*, 1966; Ito and Hotta, 1976; Ito *et al.*, 1976; Shibata *et al.*, 1973; Sparrow and van Bockxmeer, 1972). However, it is currently believed that a troponin-like system does not constitute the primary regulatory mechanism in smooth muscle and that, if it does exist, it must be secondary or complementary to another regulatory mechanism. In this regard it is interesting that Marston *et al.* (1980) have reported a dual regulatory system in smooth muscle.

The next surge of interest followed the discovery of myosin-linked regulation and, when the test for distinguishing between actin- and myosin-linked systems was applied to gizzard actomyosin, it was found that the smooth muscle system was myosin-linked (Bremel, 1974). Later reports confirmed this for chicken gizzard (Bremel *et al.*, 1977; Hartshorne *et al.*, 1977a; Ikebe *et al.*, 1977; Sobieszek and Small, 1976) and also extended the observation to different smooth muscles (Borejdo and Oplatka, 1976; Frederiksen, 1976; Mrwa and Rüegg, 1975; Takeuchi and Tonomura, 1977). However, the system in smooth muscle proved to be more complex than that in molluscan muscle, and it was discovered that as smooth muscle myosin was purified, its actin-activated ATPase activity became less and it was evident that factors in addition to actin and myosin were required for the activation of ATPase activity. These factors are generally acknowledged to be regulatory proteins. The system still retains its classification as myosin-linked, since it is thought that the activating process is centered on the myosin molecule.

Thus the system in vertebrate smooth muscle differs from that in either molluscan or vertebrate striated muscle in that a complex of homogeneous myosin and actin exhibits only negligible Mg^{2+}-ATPase activity. The primary requirement for regulation in the smooth muscle system therefore is to activate Mg^{2+}-ATPase activity, but only in the presence of Ca^{2+}. This is accepted, and the controversy that exists is centered on the identity of the activating factor. There are two theories. The most popular is that activation of ATPase activity occurs as a result of phosphorylation of the 20,000-molecular-weight myosin light chains. The second is that phosphorylation of the light chains is not involved in the regulatory process (Ebashi *et al.*, 1977; Mikawa *et al.*, 1977) but that activation is due to a system termed leiotonin. Each of these two theories will be discussed in more detail. For a more rapid comparison of the two mechanisms some of the salient points are listed in Table I.

A. Phosphorylation Theory

Sobieszek (1977a) and Bremel *et al.* (1977) first showed that the phosphorylation of the 20,000-molecular-weight light chains of chicken gizzard myosin resulted in activation of the Mg^{2+}-ATPase activity of gizzard actomyosin, and it was found also that this event was Ca^{2+}-dependent. Since these observations were made, several reports have appeared on the phosphorylation of smooth muscle myosin, and in many instances the event of phosphorylation was claimed to be a prerequisite for the activation of actomyosin ATPase activity (Aksoy *et al.*, 1976; Barron *et al.*, 1979, 1980; Chacko *et al.*, 1977; DiSalvo *et al.*, 1978; Frearson *et al.*, 1976; Gorecka *et al.*, 1976; Huszar and Bailey, 1979; Ikebe *et al.*, 1977, 1978; Sobieszek, 1977b; Sobieszek and Small, 1977). From these observations the following key points of the phosphorylation theory of regulation may be abstracted: (1) The sites of phosphorylation are the 20,000-molecular-weight light chains of the myosin molecule. (The myosin molecule consists of two heavy chains, each of about 200,000 molecular weight, and two classes of light chains, with molecular weights

TABLE I

Aspects of the Leiotonin and Phosphorylation Theories of Regulation

	Leiotonin		
	Observations	Remarks	Phosphorylation
Components	Leiotonin A (MW 80,000); leiotonin C (MW 17,000); Ca^{2+}-binding regulatory protein	Affinity for smooth muscle actin; similar to but not identical to calmodulin or troponin C; differs from calmodulin with respect to Sr^{2+} binding; does not activate phosphodiesterase; calmodulin can substitute for leiotonin C	MLCK apoenzyme (MW 77, 000–155,000) plus calmodulin regulatory protein stoichiometry 1 : 1; MLCP phosphatases I and II
Regulatory mechanism	Actin-linked, requires tropomyosin; mechanism unknown; does not involve phosphorylation of myosin; functions at approximately 1 leiotonin per 10 tropomyosins	At this stoichiometry about five leiotonin molecules per thin (2-μm) filaments	Myosin-linked, Mg^{2+}-ATPase of phosphorylated myosin activated by actin; Mg^{2+}-ATPase of dephosphorylated myosin not activated by actin

of 17,000 and 20,000, each class containing two light chains). Since 1 mole of phosphate is incorporated per 20,000-molecular-weight light chain, up to 2 moles of phosphate can be incorporated per mole of myosin. (2) Myosin light-chain kinase (MLCK) is active only in the presence of Ca^{2+} at concentrations similar to those required to initiate contraction. (3) Phosphorylation of the light chains is thought to be a prerequisite for activation of the Mg^{2+}-ATPase activity of myosin by actin. (4) A second enzyme is required to deactivate the system, namely, myosin light-chain phosphatase (MLCP).

If these facts are considered in a physiological framework, then it can be theorized that an increase in the intracellular Ca^{2+} concentration activates the MLCK which then phosphorylates myosin and initiates cross-bridge turnover (the equivalent of activation by actin of the Mg^{2+}-ATPase activity of myosin). Cross-bridge cycling will continue as long as Ca^{2+} is present, and tension development, or shortening, will result. When the Ca^{2+} level is reduced, the MLCK becomes inactive, the MLCP dephosphorylates myosin, and relaxation of muscle follows. This scheme is oversimplified, and there are several phases of the mechanism which should be considered in more detail.

1. Myosin Light-Chain Kinase

a. Properties. The initial emphasis was to identify and characterize the regulatory components. This was achieved for MLCK, and it was found by Dabrowska *et al.* (1977) that MLCK from chicken gizzard consisted of two dissimilar subunits. The molecular weights were approximately 105,000 and 17,000. It was later discovered (Dabrowska *et al.*, 1978) that the smaller subunit was identical with calmodulin (previously known as modulator protein or calcium-dependent regulator), which has been shown to bind Ca^{2+} and to regulate several Ca^{2+}-dependent processes (Cheung, 1980; Means and Dedman, 1980). The calmodulin requirement for smooth muscle MLCK was confirmed by Adelstein *et al.* (1978) using turkey gizzard and was also reported by Yazawa *et al.* (1978) for MLCK from skeletal muscle. There are now several reports of a calmodulin-dependent MLCK from a variety of tissues (Dabrowska and Hartshorne, 1978; Hathaway and Adelstein, 1979; Nairn and Perry, 1979; Waisman *et al.*, 1978; Walsh *et al.*, 1979; Yerna *et al.*, 1979). One of the variables among these reports is the size of the larger subunit. Examples are listed in Table II. It has been suggested that limited proteolysis could be partly responsible for this range of molecular weights (Adelstein *et al.*, 1978; Walsh *et al.*, 1979).

Some of the kinetic data obtained with various MLCKs are compiled in Table II. In general the V_{max} values are similar, with the exception of

TABLE II

Properties of MLCK Isolated from Different Tissues

Source	Molecular weight of apoenzyme	K_m, ATP (μM)	V_{max} (μmol min^{-1} mg^{-1})[a]	K_m, light chain (μM)	K_d, calmodulin (nM)	References
Chicken gizzard	105,000	60–70	5–15	—	—	Dabrowska *et al.*, 1977 Mrwa and Hartshorne, 1980
Turkey gizzard	125,000–130,000	—	33	—	—	Adelstein *et al.*, 1978 Conti and Adelstein, 1980
Skeletal muscle	77,000–80,000	200–400	15–30	40–50	—	Nairn and Perry, 1979; Pires and Perry, 1977
	80,000	280	4.3	24	—	Yazawa and Yagi, 1978
	155,000	—	—	—	6.6	Walsh *et al.*, 1981a
	90,000	—	5	10	~1	Blumenthal and Stull, 1980
Cardiac muscle	85,000	175	0.03	21	1.3	Walsh *et al.*, 1979
	155,000	—	—	—	2.2	Walsh *et al.*, 1981a
	94,000	—	15–22.5	—	3–5	Wolf and Hofmann, 1980
Blood platelets[b]	105,000	121	$\nless$3.1	18	—	Hathaway and Adelstein, 1979

[a] Values given for 25°C and adjusted to this temperature, if necessary, assuming a Q_{10} of 2.
[b] Data obtained using a partially purified enzyme.

one of the reports for cardiac muscle which is thought to be reduced as a result of proteolysis (Walsh *et al.*, 1979). However, the V_{max} figures were calculated using isolated myosin light chains as the substrate, and it was shown by Mrwa and Hartshorne (1980) that the rate of phosphorylation was slower when whole myosin was used. The choice of substrate may also be a factor for some of the other listed parameters, and many should be reevaluated using the native substrate, i.e., whole myosin.

Calmodulin can bind up to four molecules of Ca^{2+}, and it was shown recently that it was the Ca_4^{2+}–calmodulin complex which activates skeletal muscle MLCK (Blumenthal and Stull, 1980). The Ca_{3-4}^{2+}–calmodulin complex was shown to be required for the activation of phosphodiesterase activity (Crouch and Klee, 1980). The stoichiometry of calmodulin and the apoenzyme is 1:1 for MLCK from smooth muscle (Hartshorne *et al.*, 1980) and skeletal muscle (Blumenthal and Stull, 1980).

Some other properties of MLCK are the following: The pH dependence of chicken gizzard MLCK for the phosphorylation of isolated light chains showed a relatively broad optimum between 7 and 8 (Mrwa and Hartshorne, 1980) and this differed from the acidic pH optimum found with skeletal muscle MLCK (Pires and Perry, 1977); the Q_{10} was estimated to be about 2 (Mrwa and Hartshorne, 1980); Mg^{2+}- or Mn^{2+}-ATP was utilized as a phosphate donor (Hartshorne *et al.*, 1980), although ITP (Onishi and Watanabe, 1979), Ca^{2+}-ATP, and free ATP did not support phosphorylation (Hartshorne *et al.*, 1980); the activity of purified MLCK was not inhibited by IDP, AMP, cyclic adenosine 3′,5′-monophosphate (cyclic AMP), cyclic guanosine 3′,5′-monophosphate (cyclic GMP), or adenosine (all at concentrations up to 100 μM) (Hartshorne *et al.*, 1980).

One of the reasons for determining the kinetic parameters of MLCK is that they might serve to link the process of myosin phosphorylation with regulation in smooth muscle. To date this approach has not yielded any significant correlations. One of the reasons for this could be that MLCK is considerably faster than the cross-bridge cycling rate, calculated for arterial muscle at 37°C to be about 1 sec^{-1} (Mrwa *et al.*, 1975), and therefore phosphorylation would procede by a wide margin the tension-generating step.

b. Specificity and Site of Phosphorylation by MCLK. As stated earlier, MLCK catalyzes the incorporation of one phosphate into each of the two 20,000-molecular-weight light chains of the smooth muscle myosin molecule. The sequence near the phosphorylated serine is -Arg-Ala-Thr-Ser-Asn-Val-Phe-Ala-Met-, and this peptide is located near the N-terminus of the light chain (Jakes *et al.*, 1976). The exact substrate requirements for MLCK are not established, although it is known that the kinase exhibits a relatively high specificity. For example, MLCK will not phosphorylate histone, casein, phosphorylase b, or phosphorylase kinase (Adelstein and Eisenberg, 1980). In general the light chains from different muscles can be phosphorylated by MLCK, although there is a clear preference for the homologous system (Adelstein and Eisenberg, 1980). This tendency is more marked when whole myosin, rather than isolated light chains, is used as a substrate, and usually the phosphorylation of myosin by MLCK from a different tissue is drastically reduced or eliminated (Stull *et al.,* 1978).

An interesting feature of the sequence listed above is the presence of the arginine residue on the N-terminal side of the serine (this residue is not found in the analogous sequence of skeletal muscle light chains). It is therefore a potential substrate for cAMP-dependent protein kinase (Glass and Krebs, 1980), and indeed this was demonstrated by Noiman (1980) using isolated light chains. However, when whole myosin was tested, phosphorylation by cAMP-dependent protein kinase was not observed (Walsh *et al.,* 1981b). This again illustrates the point that the substrate characteristics of the light chain are different in the isolated and the native (intact) states.

c. Phosphorylation of MLCK by cAMP-Dependent Protein Kinase. It has been shown that the apoenzyme of MLCK is phosphorylated by cAMP-dependent protein kinase and that this phosphorylation results in a decreased affinity for calmodulin and, at limiting concentrations of calmodulin, a reduction in the MLCK rate (Adelstein *et al.,* 1978; Conti and Adelstein, 1980). More recently two sites of phosphorylation, A and B, have been identified (Adelstein *et al.,* 1981). Both are phosphorylated in the absence of calmodulin, but only the B site is labeled at saturating concentrations of calmodulin. Interestingly it is only the phosphorylation of the A site which results in a reduced affinity for calmodulin. It was proposed that this phosphorylation by cAMP-dependent protein kinase could constitute a regulatory mechanism. The theory is that, as the level of cAMP within the cell increases, an event thought to precede relaxation, the MLCK is phosphorylated and the rate of myosin phosphorylation reduced, which would facilitate relaxation. Some support for this theory has been reported, and it was shown that addition of the catalytic subunit of cAMP-dependent protein kinase resulted in a decrease in the Ca^{2+}-activated tension which was reversed on the addition of excess calmodulin (Hoar and Kerrick, 1980). Also, inhibition of the Mg^{2+}-ATPase activity (Mrwa *et al.,* 1979; Silver and DiSalvo, 1979) and of myosin phosphorylation (Silver and DiSalvo, 1979) was obtained following the addition of cAMP to smooth muscle actomyosin preparations which contained cAMP-dependent protein kinase.

2. *Myosin Light-Chain Phosphatase*

In the phosphorylation theory of regulation the mechanism can be visualized as occurring through the action of two components; the activation phase is due to MLCK, and the relaxation phase is mediated via MLCP. It is generally assumed that the MLCP is active both in the absence and presence of Ca^{2+}, but that net dephosphorylation will occur only when the MLCK is inhibited. This requires that the *in vivo* level of MLCP activity be considerably less then the MLCK activity. Although *in vivo* levels have not been assessed, MLCP activity has been detected in several smooth muscle systems (Aksoy *et al.,* 1976; Chacko *et al.,* 1977; Ikebe *et al.,* 1978; Onishi *et al.,* 1979; Sherry *et al.,* 1978), and in general phosphatase rates are lower than kinase rates.

An MLCP was first isolated from skeletal muscle and shown to be a single component of 70,000 molecular weight (Morgan *et al.,* 1976). More recently two phosphatases, I and II, have been isolated from turkey gizzard smooth muscle (Pato and Adelstein, 1980). Phosphatase I is composed of three subunits with molecular weights of 60,000, 55,000, and 38,000, and phosphatase II is a single component of 43,000 molecular weight. Both are effective in the dephosphorylation of isolated myosin light chains but differ in that only phosphatase I dephosphorylates the larger subunit of MLCK.

In general there is less information available on the MLCP system than for MLCK, and much remains to be established. For example, it would be useful to know the MLCP rates under various physiological conditions, and in this regard it is not known if the phosphatases are regulated in their native state. There is also the possibility that intact myosin exhibits substrate qualities different from those of isolated light chains, and the latter are usually used to assay for MLCP activity.

3. *Effects of Phosphorylation on the Properties of Myosin*

The most significant and publicized effect of phosphorylation is to allow the activation by actin of the Mg^{2+}-ATPase activity of myosin. This will be considered in the following section, and at this point some of the other effects of phosphorylation will be mentioned.

The effect of phosphorylation on the other ATPase activities of myosin, namely, the Ca^{2+}-ATPase, the K^+-EDTA ATPase, and the Mg^{2+}-ATPase activity of myosin alone, is not marked, although a slight effect on the last-mentioned activity has been observed (Suzuki *et al.,* 1978). However, this area has not been investigated thoroughly, and a final judgment on the effects of phosphorylation should await a more detailed study. The thick filaments formed from phosphorylated myosin have been shown to be more resistant to dissociation by Mg^{2+}-ATP than those composed of dephosphorylated myosin (Suzuki *et al.*, 1978), and it has been suggested that the filamentous state of myosin possesses a higher ATPase activity. With smooth muscle there is no evidence suggesting that the conformation is altered on phosphorylation of the light chains, although for skeletal muscle myosin this was reported and related to a change in the Ca^{2+}-binding properties of the isolated light chains (Alexis and Gratzer, 1978). Other investigators, however, found that phosphorylation did not alter the Ca^{2+}-binding properties of either cardiac (Holroyde *et al.*, 1978; Kuwayama and Yagi, 1979) or skeletal myosin (Holroyde *et al.*, 1978). It was thought that this discrepancy was due to the use of isolated light chains in one case and whole myosin in the others, but this explanation is not fully acceptable since recently some effects of phosphorylation have been observed with whole myosin (Kardami *et al.*, 1980).

It has been known for several years that both skeletal and cardiac muscle contain MLCK, although there is still no consensus on the function of myosin phosphorylation in these tissues. Pemrick (1980) has reported that for skeletal muscle myosin the effect of phosphorylation is primarily to increase the actin-binding affinity, with very little effect on the V_{max} of the actin-activated ATPase activity. It has also been suggested that in rat extensor digitorum longus muscle phosphorylation is not essential for contraction but may play a role in post-tetanic stimulation (Manning and Stull, 1979). These two findings obviously could be compatible, but this remains to be proven.

One of the intriguing problems that remains to be solved for the smooth muscle system is how phosphorylation of the light chains can affect the active sites which are located on the heavy chains. It is possible that the two sites are spatially close, and any information pertinent to this would be most useful. Also, it is not established whether the dephosphorylated light chain inhibits the ATPase activity, which is relieved by phosphorylation, or if the dephosphorylated light chain is inert and the phosphorylated light chain activates the dormant state. Similarly it is not known what the relationship is between the two heads of myosin and the two phosphorylatable light chains. Is each head independent or are there cooperative interactions between the two?

4. *A Critique of the Phosphorylation Theory of Regulation*

The observation that forms the basis of the phosphorylation theory is that myosin phosphorylation is required for the actin-activated ATPase activity. As mentioned previously, this relationship has been confirmed with actomyosin from several different smooth muscles and has also been extended with various fiber preparations to a correlation of tension with myosin phosphorylation (Barron *et al.*, 1979, 1980; Driska *et al.*, 1981; Hoar *et al.*, 1979; Janis and Gualteri, 1978). Thus there is a wealth of information which indicates that the phosphorylation of myosin is associated with the contractile process, but these studies cannot eliminate the involvement of other factors. This is primarily because none of the systems used were adequately defined with respect to their protein components, and thus effects due to an unidentified factor could not be eliminated. Ebashi and his colleagues (Section IV,B) have claimed for several years that myosin phosphorylation is not involved in the regulatory process in smooth muscle but that regulation is due rather to a system based on the thin filaments, termed leiotonin. Clearly one of the priorities for future research is to establish the relationship between myosin phosphorylation and the extent of actin-activated ATPase activity under carefully defined conditions. Until this is accomplished, we can only point out some of the inconsistencies associated with the phosphorylation theory of regulation.

Dillon *et al.* (1981), using strips of muscle from swine carotid arteries, found that phosphorylation of myosin preceded tension development but that the extent of myosin phosphorylation subsequently declined significantly while the load-bearing capacity was maintained. These results were interpreted to suggest that phosphorylation was closely correlated with cross-bridge cycling (to explain the initial rise in tension) and also that following dephosphorylation an attached, noncycling cross-bridge was formed (to explain the maintenance of the load-bearing capacity). The idea of noncycling attached cross-bridges is not new, and they have been mentioned previously (Siegman *et al.*, 1976a,b). However, what initiates the formation of these attached cross-bridges and what brings about their eventual dissociation are not known. If the latter is Ca^{2+}-dependent (and this is a reasonable expectation), then a second Ca^{2+}-dependent mechanism (distinct from the MLCK system) must exist, and this would not be consistent with the simplest interpretation of the phosphorylation theory. In a study using tracheal smooth muscle deLanerolle and Stull (1980) found that myosin phosphorylation coincided temporally with the increase in isometric tension, but that during atropine-induced relaxation the decrease in tension was faster than the drop in the level of myosin phosphorylation. Again this suggests that a process other than the dephosphorylation of myosin might be involved in the relaxation process. Murray and England (1980), using intact pig aortic strips, found that preincubation with EGTA followed by the addition of noradrenaline elicited an increase in tension but no increase in the extent of myosin phosphorylation. Therefore it was suggested that there must be another mechanism capable of regulating contractile activity in this smooth muscle.

The above discussion considers some of the results which suggest that phosphorylation alone may not be adequate to account for the regulation of smooth muscle activity. On the other hand, there are also several approaches which strongly support the phosphorylation theory (including the use of phenothiazine derivatives and thiophosphorylation; for a discussion, see Hartshorne and Mrwa, 1981), and therefore the overall impression is that phosphorylation of myosin is an essential component of the regulatory mechanism in smooth muscle, and that in addition other factors might be involved. Whether the latter constitute independent or complementary mechanisms remains to be established.

B. Leiotonin

The alternative system that has been proposed by Ebashi and colleagues as the regulatory mechanism in smooth muscle is leiotonin. This is similar to the phosphorylation theory in that it effects activation of actomyosin ATPase activity, but it differs in several other respects (Table I). The most important difference is that leiotonin is thought to function independently of myosin phosphorylation which is not considered to be involved in the regulation of smooth muscle activity (Ebashi *et al.*, 1977; Mikawa *et al.*, 1977). Another point of distinction is that leiotonin is thought to be located on the thin filaments and, unlike the phosphorylation system, is not myosin-linked. The experiments of Mikawa (1979) support this contention, and it was found that glutaraldehyde could "freeze" the thin filaments of smooth muscle into either an active or an inactive state. In addition it was shown that the leiotonin–actin complex demonstrated a strong affinity for Ca^{2+}, which was reduced when leiotonin was removed (Hirata *et al.*, 1980).

Leiotonin is composed of two subunits, A (MW ~ 80,000) and C (MW ~ 17,000) (Mikawa *et al.*, 1978; Nonomura and Ebashi, 1980). Leiotonin A shows an affinity for smooth muscle actin but not for tropomyosin, although this protein is an essential component for the leiotonin mechanism. The amount of leiotonin in smooth muscle is quite low, and it has been estimated that leiotonin is functional at a ratio of about 1 molecule to 10 tropomyosin molecules (Ebashi *et al.*, 1977; Nonomura and Ebashi, 1980). This relatively low stoichiometry with respect to the other thin-filament components (actin and tropomyosin) argues against a structural role for leiotonin (analogous to that of troponin in skeletal muscle) and might indicate that it functions via some enzymic pathway.

Leiotonin C is an acidic protein, similar but not identical to calmodulin (Mikawa *et al.,* 1978). It lacks trimethyllysine, which is found in vertebrate calmodulins, and cannot activate phosphodiesterase activity (Nonomura and Ebashi, 1980). However, calmodulin can substitute for leiotonin C in the activation of smooth muscle actomyosin (Mikawa *et al.,* 1978). When leiotonin C is compared to troponin C, some similarities are evident, including electrophoretic mobilities in the presence and absence of Ca^{2+} (Nonomura and Ebashi, 1980), thus leading to the suggestion that both leiotonin C and troponin C might be evolutionary descendants of calmodulin.

Since calmodulin can substitute for leiotonin C, one of the questions which arose was, which of the two is the *in vivo* regulatory protein? This problem was solved by a consideration of the Sr^{2+} dependency of superprecipitation. It was found (Hirata *et al.,* 1980) that, in a reconstituted system containing leiotonin A plus calmodulin, the Sr^{2+} curve for superprecipitation was shifted to lower concentrations of Sr^{2+} than those observed with the native system. The key point is that leiotonin C binds Sr^{2+} less effectively than calmodulin.

Obviously the two alternative mechanisms of regulation in smooth muscle must be understood before their relationship to each other can be established. The most important difference at this time is the lack of dependence of the leiotonin system on myosin phosphorylation. The mode of action of leiotonin also deserves high priority for future research, and until more information is available the relative merits of the two systems cannot be evaluated.

C. Other Regulatory Mechanisms

In the above discussion emphasis was placed on either the phosphorylation theory or the leiotonin system, and attempts were made to point out deviations from the basic hypotheses. However, there are a few reports on regulatory effects that do not fit with either system, and these are considered here. The fact that they do not agree with preconceived ideas may merely point to our ignorance but, on the other hand, they could represent novel regulatory systems.

Seidel (1979) made the interesting observation that gizzard myosin modified by sulfhydryl reagents no longer required phosphorylation for the activation of ATPase activity by actin. However, it was also found using the modified myosin that activation by actin did not occur with pure skeletal muscle actin and was observed only when a crude tropomyosin fraction was added. Purified tropomyosin was not effective, and these results therefore point to the existence of an additional activating component. It is possible that the unknown factor could be leiotonin, although Ebashi *et al.* (1975) state that this system is effective only when smooth muscle actin is used.

Marston *et al.* (1980), using actomyosin from pig aorta and from turkey gizzard, found evidence of both actin- and myosin-linked regulation. A thin-filament preparation was isolated from pig aorta and, although many of the protein components were unidentified, some evidence was presented to suggest the presence of troponin I- and troponin C-like proteins. The Ca^{2+}-dependent activation of the Mg^{2+}-ATPase activity of skeletal muscle myosin by the thin filaments, and the Ca^{2+}-binding properties of the thin filaments, were consistent with the presence of a troponin-like mechanism. The amount of troponin-like components was slightly less than the amount expected using skeletal muscle stoichiometry. However, the most significant difference between aorta thin filaments and classic troponin-containing filaments was that the amount of Ca^{2+} bound was significantly reduced in the absence of ATP. This is surprising but is presumably related to a subsequent finding by Marston and Walters (1980) that thin-filament preparations contain enzymes which phosphorylate and dephosphorylate a specific protein. The latter is a basic protein (MW 20,000), and in its phosphorylated form it increases the calcium-binding affinity of thin filaments by about 10-fold. Thus, despite some similarities to the troponin–tropomyosin complex, the pig aorta system is unique and apparently is subject to another as yet unidentified regulatory kinase–phosphatase combination. One reason why this system had escaped previous identification is that the washing procedures used in various thin-filament preparations vary considerably and some components of the complete system could easily be lost.

In the above discussion we considered factors that might be required in addition to the actomyosin–MLCK–MLCP systems; however, there remains a regulatory component that might be associated with the myosin molecule itself. This is the possibility of regulation via Ca^{2+} binding to the myosin light chains (similar to the molluscan system). If it is assumed that the phosphorylation theory of regulation is functional, then in its simplest concept the role of Ca^{2+} is directed only toward one site, calmodulin, and Ca^{2+} functions by activating MLCK (Ikebe *et al.*, 1978; Sherry *et al.*, 1978; Small and Sobieszek, 1977a,b). A slightly more complex hypothesis is that Ca^{2+} binding to myosin, in addition to phosphorylation, is a requirement for regulation (Chacko *et al.*, 1977; Rees and Frederiksen, 1981). The experimental observations described in support of the latter possibility were that the actin-activated ATPase activity of phosphorylated myosin was inhibited on the removal of Ca^{2+}. In a physiological context therefore the removal of Ca^{2+} from myosin would reduce cross-bridge cycling and might be implicated in the relaxation process. However, at the moment this possibility is purely speculative, as there is considerable controversy regarding the role of the Ca^{2+}–myosin interaction. One problem is the amount of Ca^{2+} expected to be bound to myosin under physiological conditions. It is known that Mg^{2+} competes for the Ca^{2+}-binding sites, and thus the concentration of free Mg^{2+} is an important consideration. For example, with bovine aorta myosin the amount of Ca^{2+} bound in the presence of 8 mM Mg^{2+} was insignificant (Hirata *et al.,* 1980), and for rabbit skeletal myosin at 1×10^{-5} M Ca^{2+} and 1 mM Mg^{2+} only about 50% of the sites were occupied by Ca^{2+} (Watterson *et al.,* 1979). Thus, before any conclusions can be reached, it will be necessary to determine the extent of Ca^{2+} binding to myosin under conditions approximating the *in vivo* state.

The final point to consider in this section is the role of tropomyosin. If the existence of a troponin-like mechanism is confirmed, then the function of tropomyosin will be obvious. If, however, this is not forthcoming, then the role of tropomyosin will be less clear, and at present there is no evidence for either the phosphorylation or the leiotonin theory that tropomyosin forms a stoichiometric complex with one of the regulatory components. It is, however, a consistent observation (Chacko *et al.,* 1977; Hartshorne *et al.,* 1977b; Nonomura and Ebashi, 1980; Sobieszek and Small, 1977) that tropomyosin is required for full activation of the ATPase activity of smooth muscle actomyosin. Possibly, tropomyosin is involved in the potentiation (along the thin filament) of some effect induced by a separate mechanism, and this could fit either the phosphorylation or the leiotonin theory.

D. Summary of the Regulatory Mechanism in Smooth Muscle

It is generally accepted that the activity of the contractile apparatus in smooth muscle is regulated by the intracellular concentration of Ca^{2+}. The increase or decrease of available Ca^{2+} is sensed by some component associated with the contractile proteins, with the result that cross-bridge interactions with actin are either promoted or dissociated, respectively. The identity of this sensing system, i.e., the regulatory proteins, is the subject of controversy, and two major alternatives have been suggested. The most widely accepted theory is that regulation is achieved through the phosphorylation and dephosphorylation of the myosin molecule and therefore involves both a kinase and a phosphatase system. The alternative viewpoint is that regulation is achieved by a system termed leiotonin.

There is a considerable amount of experimental support for the phosphorylation theory, and one is forced to conclude that the phosphorylation of myosin forms an essential component of the regulatory mechanism in smooth muscle. It has been shown, for example, that contraction does not occur in the absence of myosin phosphorylation, that phosphorylation precedes tension development, and that phosphorylation appears to regulate the cross-bridge cycling rate. On the other hand, there appears to be no simple relationship between myosin phosphorylation and tension development which is consistent with all the observed data, and therefore it appears that in addition to phosphorylation other factors should be considered.

The leiotonin system is quite different. It does not function via the phosphorylation of myosin, and its site of action appears to be associated with the thin filament. The major deficiency of the leiotonin theory is that its mechanism is not known, and this must be established before it can be decided whether the two major theories of regulation in smooth muscle are really independent or form some complementary mode of regulation.

ACKNOWLEDGMENT

The author's research was supported by grant HL 23615 from the National Institutes of Health.

REFERENCES

Adelstein, R. S., and Eisenberg, E. (1980). Regulation and kinetics of the actin-myosin-ATP interaction. *Annu. Rev. Biochem.* **49,** 921-956.

Adelstein, R. S., Conti, M. A., Hathaway, D. R., and Klee, C. B. (1978). Phosphorylation of smooth muscle myosin light chain kinase by the catalytic subunit of adenosine 3′:5′-monophosphate-dependent protein kinase. *J. Biol. Chem.* **253,** 8347-8350.

Adelstein, R. S., Pato, M. D., and Conti, M. A. (1981). The role of phosphorylation in regulating contractile proteins. *Adv. Cyclic Nucleotide Res.* **14** (in press).

Aksoy, M. O., Williams, D., Sharkey, E. M., and Hartshorne, D. J. (1976). A relationship between Ca^{2+} sensitivity and phosphorylation of gizzard actomyosin. *Biochem. Biophys. Res. Commun.* **69,** 35-41.

Alexis, M. N., and Gratzer, W. B. (1978). Interaction of skeletal myosin light chains with calcium ions. *Biochemistry* **17,** 2319-2325.

Ashton, F. T., Somlyo, A. V., and Somlyo, A. P. (1975). The contractile apparatus of vascular smooth muscle: Intermediate high voltage stereo electron microscopy. *J. Mol. Biol.* **98,** 17-29.

Barron, J. T., Bárány, M., and Bárány, K. (1979). Phosphorylation of the 20,000-dalton light chain of myosin of intact arterial smooth muscle in rest and in contraction. *J. Biol. Chem.* **254,** 4954–4956.

Barron, J. T., Bárány, M., Bárány, K., and Storti, R. V. (1980). Reversible phosphorylation and dephosphorylation of the 20,000-dalton light chain of myosin during the contraction-relaxation-contraction cycle of arterial smooth muscle. *J. Biol. Chem.* **255,** 6238–6244.

Blumenthal, D. K., and Stull, J. T. (1980). Activation of skeletal muscle myosin light chain kinase by calcium (2+) and calmodulin. *Biochemistry* **19,** 5608–5614.

Borejdo, J., and Oplatka, A. (1976). Evidence for myosin-linked regulation in guinea pig taenia coli muscle. *Pfluegers Arch.* **366,** 177–184.

Bremel, R. D. (1974). Myosin linked calcium regulation in vertebrate smooth muscle. *Nature (London)* **252,** 405–407.

Bremel, R. D., Sobieszek, A., and Small, J. V. (1977). Regulation of actin-myosin interaction in vertebrate smooth muscle. *In* "The Biochemistry of Smooth Muscle" (N. L. Stephens, ed.), pp. 533–549. University Park Press, Baltimore, Maryland.

Carsten, M. E. (1971). Uterine smooth muscle: Troponin. *Arch. Biochem. Biophys.* **147,** 353–357.

Chacko, S., Conti, M. A., and Adelstein, R. S. (1977). Effect of phosphorylation of smooth muscle myosin on actin activation and Ca^{2+} regulation. *Proc. Natl. Acad. Sci. U.S.A.* **74,** 129–133.

Chantler, P. D., Sellers, J. R., and Szent-Györgyi, A. G. (1981). Cooperativity in scallop myosin. *Biochemistry* **20,** 210–216.

Cheung, W. Y. (1980). Calmodulin plays a pivotal role in cellular regulation. *Science* **207,** 19–27.

Cohen, D. M., and Murphy, R. A. (1978). Differences in cellular contractile protein contents among porcine smooth muscles: Evidence for variation in the contractile system. *J. Gen. Physiol.* **72,** 369–380.

Conti, M. A., and Adelstein, R. S. (1980). Phosphorylation by cyclic adenosine 3′:5′-monophosphate-dependent protein kinase regulates myosin light chain kinase. *Fed. Proc. Fed. Am. Soc. Exp. Biol.* **39,** 1569–1573.

Crouch, T. H., and Klee, C. B. (1980). Positive cooperative binding of calcium to bovine brain calmodulin. *Biochemistry* **19,** 3692–3698.

Dabrowska, A., Aromatorio, D., Sherry, J. M. F., and Hartshorne, D. J. (1977). Composition of the myosin light chain kinase from chicken gizzard. *Biochem. Biophys. Res. Commun.* **78,** 1263–1272.

Dabrowska, R., and Hartshorne, D. J. (1978). A Ca^{2+}- and modulator-dependent myosin light chain kinase from non-muscle cells. *Biochem. Biophys. Res. Commun.* **85,** 1352–1359.

Dabrowska, R., Sherry, J. M. F., Aromatorio, D. K., and Hartshorne, D. J. (1978). Modulator protein as a component of the myosin light chain kinase from chicken gizzard. *Biochemistry* **17,** 253–258.

deLanerolle, P., and Stull, J. T. (1980). Myosin phosphorylation during contraction and relaxation of tracheal smooth muscle. *J. Biol. Chem.* **255,** 9993–10,000.

Devine, C. E., and Somlyo, A. P. (1971). Thick filaments in vascular smooth muscle. *J. Cell Biol.* **49,** 636–649.

Dillon, P. F., Aksoy, M. O., Driska, S. P., and Murphy, R. A. (1981). Myosin phosphorylation and the cross-bridge cycle in arterial smooth muscle. *Science* **211,** 495–497.

DiSalvo, J., Gruenstein, E., and Silver, P. (1978). Ca^{2+} dependent phosphorylation of bovine aortic actomyosin. *Proc. Soc. Exp. Biol. Med.* **158,** 410–414.

Driska, S., and Hartshorne, D. J. (1975). The contractile proteins of smooth muscle:

Properties and components of a Ca^{2+}-sensitive actomyosin from chicken gizzard. *Arch. Biochem. Biophys.* **167,** 203-212.
Driska, S., Aksoy, M. O., and Murphy, R. A. (1981). Myosin light chain phosphorylation associated with contraction in arterial smooth muscle. *Am. J. Physiol.* **240,** C222-C233.
Ebashi, S. (1963). Third component participating in the superprecipitation of "natural actomyosin." *Nature (London)* **200,** 1010.
Ebashi, S., Iwakura, H., Nakajima, H., Nakamura, R., and Ooi, Y. (1966). New structural proteins from dog heart and chicken gizzard. *Biochem. Z.* **345,** 201-211.
Ebashi, S., Kodama, A., and Ebashi, F. (1968). Troponin I: Preparation and physiological function. *J. Biochem. (Tokyo)* **64,** 465-477.
Ebashi, S., Toyo-oka, R., and Nonomura, Y. (1975). Gizzard troponin. *J. Biochem. (Tokyo)* **78,** 859-861.
Ebashi, S., Mikawa, T., Hirata, M., Toyo-oka, T., and Nonomura, Y. (1977). Regulatory proteins of smooth muscle. *In* "Excitation-Contraction Coupling in Smooth Muscle" (R. Casteels, T. Godfraind, and J. C. Rüegg, eds.) pp. 325-334. Elsevier/North-Holland Biomedical Press, New York.
Frearson, N., Focant, B. W. W., and Perry, S. V. (1976). Phosphorylation of a light chain component of myosin from smooth muscle. *FEBS Lett.* **63,** 27-32.
Frederiksen, D. W. (1976). Myosin-mediated Ca^{++}-regulation of actomyosin-adenosinetriphosphatase from porcine aorta. *Proc. Natl. Acad. Sci. U.S.A.* **73,** 2706-2710.
Glass, D. B., and Krebs, E. G. (1980). Protein phosphorylation catalyzed by cyclic AMP-dependent and cyclic GMP-dependent protein kinases. *Annu. Rev. Pharmacol. Toxicol.* **20,** 363-388.
Gorecka, A., Aksoy, M. O., and Hartshorne, D. J. (1976). The effect of phosphorylation of gizzard myosin on actin activation. *Biochem. Biophys. Res. Commun.* **71,** 325-331.
Hartshorne, D. J., and Gorecka, A. (1980). The biochemistry of the contractile proteins of smooth muscle. *In* "Handbook of Physiology" (D. F. Bohr, A. P. Somlyo, and H. V. Sparks, eds.), Sec. 2, Vol. II, pp. 93-120. Am. Physiol. Soc. Bethesda, Maryland.
Hartshorne, D. J., and Mrwa, U. (1981). Regulation of smooth muscle actomyosin. *Blood Vessels* (in press).
Hartshorne, D. J., Abrams, L., Aksoy, M. O., Dabrowska, R., Driska, S., and Sharkey, E. M. (1977a). Molecular basis for the regulation of smooth muscle actomyosin. *In* "The Biochemistry of Smooth Muscle" (N. L. Stephens, ed.), pp. 513-532. University Park Press, Baltimore, Maryland.
Hartshorne, D. J., Gorecka, A., and Aksoy, M. O. (1977b). Aspects of the regulatory mechanism in smooth muscle. *In* "Excitation-Contraction Coupling in Smooth Muscle" (R. Casteels, T. Godfraind, and J. C. Ruegg, eds.), pp. 377-384. Elsevier/North-Holland Biomedical Press, New York.
Hartshorne, D. J., Siemankowski, R. F., and Aksoy, M. O. (1980). Ca regulation in smooth muscle and phosphorylation: Some properties of the myosin light kinase. *In* "Muscle Contraction. Its Regulatory Mechanisms" (S. Ebashi, K. Maruyama, and M. Endo, eds.), pp. 287-301, Springer-Verlag, New York and Berlin. Japan Scientific Societies Press, Tokyo.
Hathaway, D. R., and Adelstein, R. S. (1979). Human platelet myosin light chain kinase requires the calcium-binding protein calmodulin for activity. *Proc. Natl. Acad. Sci. U.S.A.* **76,** 1653-1657.
Hinssen, H., D'Haese, J., Small, J. V., and Sobieszek, A. (1978). Mode of filament assembly of myosins from muscle and non-muscle cells. *J. Ultrastruct. Res.* **64,** 282-302.
Hirata, M., Mikawa, T., Nonomura, Y., and Ebashi, S. (1980). Ca^{2+} regulation in vascular smooth muscle. II. Ca^{2+} binding of aorta leiotonin. *J. Biochem. (Tokyo)* **87,** 369-378.

Hoar, P. E., and Kerrick, W. G. L. (1980). Catalytic subunit of c-AMP dependent protein kinase: Effect on contraction of functionally skinned muscle fibers. *Fed. Proc. Fed. Am. Soc. Exp. Biol.* **34,** 1817.

Hoar, P. E., Kerrick, W. G. L., and Cassidy, P. A. (1979). Chicken gizzard: Relation between calcium-activated phosphorylation and contraction. *Science* **204,** 503-506.

Holroyde, M. J., Potter, J. D., and Solaro, R. J. (1978). The calcium binding properties of phosphorylated and unphosphorylated cardiac and skeletal myosins. *J. Biol. Chem.* **254,** 6478-6482.

Huszar, G., and Bailey, P. (1979). Relationship between actin-myosin interaction and myosin light chain phosphorylation in human placental smooth muscle. *Am. J. Obstet. Gynecol.* **135,** 718-726.

Huxley, A. F. (1957). Muscle structure and theories of contraction. *Prog. Biophys. Mol. Biol.* **7,** 257-318.

Huxley, A. F., and Niedergerke, R. (1954). Structural changes in muscle during contraction. *Nature (London)* **173,** 971-973.

Huxley, H. E., and Hanson, J. (1954). Changes in the cross-striations of muscle during contraction and stretch and their structural interpretation. *Nature (London)* **173,** 973-976.

Ikebe, M., Onishi, H., and Watanabe, S. (1977). Phosphorylation and dephosphorylation of a light chain of the chicken gizzard myosin molecule. *J. Biochem. (Tokyo)* **82,** 299-302.

I'kebe, M., Aiba, T., Onishi, H., and Watanabe, S. (1978). Calcium sensitivity of contractile proteins from chicken gizzard muscle. *J. Biochem. (Tokyo)* **83,** 1643-1655.

Ito, N., and Hotta, K. (1976). Regulatory protein of bovine tracheal smooth muscle. *J. Biochem. (Tokyo)* **80,** 401-403.

Ito, N., Takagi, T., and Hotta, K. (1976). Regulatory protein of vascular smooth muscle. *J. Biochem. (Tokyo)* **80,** 899-901.

Jakes, R., Northrop, F., and Kendrick-Jones, J. (1976). Calcium binding regions of myosin "regulatory" light chains. *FEBS Lett.* **70,** 229-234.

Janis, R. A., and Gualteri, R. T. (1978). Contraction of intact smooth muscle is associated with the phosphorylation of a 20,000-dalton protein. *Physiologist* **21,** 59.

Johansson, B., and Somlyo, A. P. (1980). Electrophysiology and excitation-contraction coupling. *In* "Handbook of Physiology" (D. F. Bohr, A. P. Somlyo, and H. V. Sparks, eds.), Sect. 2, Vol. II, pp. 301-323. Am. Physiol. Soc., Bethesda, Maryland.

Kardami, E., Alexis, M., de la Paz, P., and Gratzer, W. (1980). Phosphorylation and the binding of calcium and magnesium to skeletal myosin. *Eur. J. Biochem.* **110,** 153-160.

Kelley R. E., and Rice, R. V. (1968). Localization of myosin filaments in smooth muscle. *J. Cell Biol.* **37,** 105-116.

Kendrick-Jones, J., Lehman, W., and Szent-Györgyi, A. G. (1970). Regulation in molluscan muscles. *J. Mol. Biol.* **54,** 313-326.

Kendrick-Jones, J., Szentkiralyi, E. M., and Szent-Györgyi, A. G. (1976). Regulatory light chains in myosins. *J. Mol. Biol.* **104,** 747-775.

Kuwayama, H., and Yagi, K. (1979). Ca^{2+} binding of pig cardiac myosin subfragment-1 and g_2 light chain. *J. Biochem. (Tokyo)* **85,** 1245-1255.

Lehman, W., and Szent-Györgyi, A. G. (1975). Regulation of muscular contraction: Distribution of actin control and myosin control in the animal kingdom. *J. Gen. Physiol.* **66,** 1-30.

Lehman, W.,.Kendrick-Jones, J., and Szent-Györgyi, A. G. (1972). Myosin-linked regulatory systems: Comparative studies. *Cold Spring Harbor Symp. Quant. Biol.* **37,** 319-330.

Manning, D. R., and Stull, J. T. (1979). Myosin light chain phosphorylation and phosphorylase *a* activity in rat extensor digitorum longus muscle. *Biochem. Biophys. Res. Commun.* **90,** 164–170.

Marston, S. B., and Walters, M. (1980). Phosphorylation of thin filaments regulates vascular smooth muscle actomyosin ATPase activity. *Cell Biol. Int. Rep.* **4,** 799.

Marston, S. B., Trevett, R. M., and Walters, M. (1980). Calcium ion-regulated thin filaments from vascular smooth muscle. *Biochem. J.* **185,** 355–365.

Means, A. R., and Dedman, J. R. (1980). Calmodulin-An intracellular calcium receptor. *Nature (London)* **285,** 73–77.

Mikawa, T. (1979) Freezing of the calcium-regulated structures of gizzard thin filaments by glutaraldehyde. *J. Biochem. (Tokyo)* **85,** 879–881.

Mikawa, T., Nonomura, Y., and Ebashi, S. (1977). Does phosphorylation of myosin light chain have a direct relation to regulation in smooth muscle? *J. Biochem. (Tokyo)* **82,** 1789–1791.

Mikawa, T., Nonomura, Y., Hirata, M., Ebashi, S., and Kakiuchi, S. (1978). Involvement of an acidic protein in regulation of smooth muscle contraction by the tropomyosin-leiotonin system. *J. Biochem. (Tokyo)* **84,** 1633–1636.

Morgan, M., Perry, S. V., and Ottaway, J. (1976). Myosin light-chain phosphatase. *Biochem. J.* **157,** 687–697.

Mrwa, U., and Hartshorne, D. J. (1980). Phosphorylation of smooth muscle myosin and myosin light chains. *Fed. Proc. Fed. Am. Soc. Exp. Biol.* **39,** 1564–1568.

Mrwa, U., and Rüegg, J. C. (1975). Myosin-linked calcium regulation in vascular smooth muscle. *FEBS Lett.* **60,** 81–84.

Mrwa, U., Paul, R. J., Kreye, V. A. W., and Rüegg, J. C. (1975). The contractile mechanism of vascular smooth muscle. *Colloq.—Inst. Natl. Sante Rech. Med.* **50,** 319–326.

Mrwa, U., Troschka, M., and Rüegg, J. C. (1979). Cyclic AMP-dependent inhibition of smooth muscle actomyosin. *FEBS Lett.* **107,** 371–374.

Murphy, R. A. (1979). Filament organization and contractile function in vertebrate smooth muscle. *Annu. Rev. Physiol.* **41,** 737–748.

Murphy, R. A., Herlihy, J. T., and Megerman, J. (1974). Force-generating capacity and contractile protein content of arterial smooth muscle. *J. Gen. Physiol.* **64,** 691–705.

Murray, K. J., and England, P. J. (1980). Contraction in intact pig aortic strips is not always associated with phosphorylation of myosin light chains. *Biochem. J.* **192,** 967–970.

Murray, J. M., and Weber, A. (1974). The cooperative action of muscle proteins. *Sci. Am.* **230,** 58–71.

Nairn, A. C., and Perry, S. V. (1979). Calmodulin and myosin light-chain kinase of rabbit fast skeletal muscle. *Biochem. J.* **179,** 89–97.

Noiman, E. S. (1980). Phosphorylation of smooth muscle myosin light chains by cAMP-dependent protein kinase. *J. Biol. Chem.* **255,** 11067–11070.

Nonomura, Y. (1968). Myofilaments in smooth muscle of guinea pig's taenia coli. *J. Cell Biol.* **39,** 741–745.

Nonomura, Y., and Ebashi, S. (1980). Calcium regulatory mechanism in vertebrate smooth muscle. *Biomed. Res.* **1,** 1–14.

Nonomura, Y., Mikawa, T., and Ebashi, S. (1980). Ca^{2+} sensitive thin filament from chicken gizzard smooth muscle. *Proc. Jpn. Acad. Ser. B* **56,** 178–183.

Onishi, H., and Watanabe, S. (1979). Calcium regulation in chicken gizzard muscle and inosine triphosphate-induced superprecipitation of skeletal acto-gizzard myosin. *J. Biochem. (Tokyo)* **86,** 569–573.

Onishi, H., Iijima, S., Anzai, H., and Watanabe, S. (1979). The possible role of myosin light-chain phosphatase in relaxation of chicken gizzard muscle. *J. Biochem. (Tokyo)* **86,** 1283–1290.

Pato, M. D., and Adelstein, R. S. (1980). Dephosphorylation of the 20,000-dalton light chain of myosin by two different phosphatases from smooth muscle. *J. Biol. Chem.* **255,** 6535-6538.

Pemrick, S. J. (1980). The phosphorylated L_2 light chain of skeletal myosin is a modifier of the actomyosin ATPase. *J. Biol. Chem.* **255,** 8836-8841.

Pires, E. M. V., and Perry, S. V. (1977). Purification and properties of myosin light-chain kinase from fast skeletal muscle. *Biochem. J.* **167,** 137-146.

Rees, D. D., and Frederiksen, D. W. (!981). Calcium regulation of porcine aortic myosin. *J. Biol. Chem.* **256,** 357-364.

Scholey, J. M., Taylor, K. A., and Kendrick-Jones, J. (1980). Regulation of non-muscle myosin assembly by calmodulin-dependent light chain kinase. *Nature (London)* **287,** 233-235.

Schollmeyer, J. E., Furcht, L. T., Goll, D. E. Robson, R. M., and Stromer, M. H. (1976). Localization of contractile proteins in smooth muscle cells and in normal and transformed fibroblasts. *Cold Spring Harbor Conf. Cell Proliferation* **3,** [Book A], 361-388.

Seidel, J. C. (1979). Activation by actin of ATPase activity of chemically modified gizzard myosin without phosphorylation. *Biochem. Biophys. Res. Commun.* **89,** 958-964.

Sherry, J. M. F., Gorecka, A., Aksoy, M. O., Dabrowska, R., and Hartshorne, D. J. (1978). Roles of calcium and phosphorylation in the regulation of the activity of gizzard myosin. *Biochemistry* **17,** 4411-4418.

Shibata, N., Yamagami, R., Yoneda, S., Akagami, H., Takeuchi, K. Tanaka, K., and Okamura, Y. (1973). Identification of myosin A, actin, and native tropomyosin constitution of arterial contractile protein (myosin B) and their characteristics. *Jpn. Circ. J.* **37,** 229-252.

Siegman, M. J., Butler, T. M., Mooers, S. U., and Davies, R. E. (1976a). Calcium-dependent resistance to stretch and stress relaxation in resting smooth muscles. *Am. J. Physiol.* **231,** 1501-1508.

Siegman, M. J., Butler, T. M., Mooers, S. U., and Davies, R. E. (1976b). Cross-bridge attachment, resistance to stretch, and viscoelasticity in resting mammalian smooth muscle. *Science* **191,** 383-385.

Silver, P. J., and DiSalvo, J. (1979). Adenosine 3′:5′-monophosphate-mediated inhibition of myosin light chain phosphorylation in bovine aortic actomyosin. *J. Biol. Chem.* **254,** 9951-9954.

Small, J. V. (1977). Studies on isolated smooth muscle cells: The contractile apparatus. *J. Cell Sci.* **24,** 327-349.

Small, J. V., and Sobieszek, A. (1977a). Ca regulation of mammalian smooth muscle actomyosin via a kinase-phosphatase-dependent phosphorylation and dephosphorylation of the 20,000-M_r light chain of myosin. *Eur. J. Biochem.* **76,** 521-530.

Small, J. V., and Sobieszek, A. (1977b). Myosin phosphorylation and Ca regulation in vertebrate smooth muscle. *In* "Excitation-Contraction Coupling in Smooth Muscle" (R. Casteels, T. Godfraind, and J. C. Rüegg, eds.), pp. 385-393. Elsevier/North-Holland Biomedical Press, Amsterdam.

Small, J. V., and Sobieszek, A. (1980). The contractile apparatus of smooth muscle. *Int. Rev. Cytol.* **64,** 241-306.

Sobieszek, A. (1977a). Vertebrate smooth muscle myosin: Enzymatic and structural properties. *In* "The Biochemistry of Smooth Muscle" (N. L. Stephens, ed.), pp. 413-443. University Park Press, Baltimore, Maryland.

Sobieszek, A. (1977b). Ca-linked phosphorylation of a light chain of vertebrate smooth-muscle myosin. *Eur. J. Biochem.* **73,** 477-483.

Sobieszek, A., and Small, J. V. (1976). Myosin-linked calcium regulation in vertebrate smooth muscle. *J. Mol. Biol.* **102,** 75-92.

Sobieszek, A., and Small, J. V. (1977). Regulation of the actin–myosin interaction in vertebrate smooth muscle: Activation via a myosin light-chain kinase and the effect of tropomyosin. *J. Mol. Biol.* **112,** 559–576.

Somlyo, A. V. (1980). Ultrastructure of vascular smooth muscle. *In* "Handbook of Physiology" (D. F. Bohr, A. P. Somlyo, and H. V. Sparks, eds), Sect. 2, Vol. II, pp. 33–67. Am. Physiol. Soc., Bethesda, Maryland.

Sparrow, M. P., and van Bockxmeer, F. M. (1972). Arterial tropomyosin and a relaxing protein fraction from vascular smooth muscle: Comparison with skeletal tropomyosin and troponin. *J. Biochem. (Tokyo)* **72,** 1075–1080.

Stull, J. T., Blumenthal, D. K., deLanerolle, P., High, C. W., and Manning, D. R. (1978). Phosphorylation and regulation of contractile proteins. *Adv. Pharmacol. Ther.* **3,** 171–180.

Suzuki, H., Onishi, H., Takahashi, K., and Watanabe, S. (1978). Structure and function of chicken gizzard myosin. *J. Biochem. (Tokyo)* **84,** 1529–1542.

Szent-Györgyi, A. G., Szentkiralyi, E. M., and Kendrick-Jones, J. (1973). The light chains of scallop myosin as regulatory subunits. *J. Mol. Biol.* **74,** 179–203.

Takeuchi, K., and Tonomura, Y. (1977). Kinetic and regulatory properties of myosin adenosinetriphosphatase purified from arterial smooth muscle. *J. Biochem. (Tokyo)* **82,** 813–833.

van Breemen, C., Aaronson, P., Loutzenhiser, R., and Meisheri, K. (1980). Ca^{2+} movements in smooth muscle. *Chest* **78,** Suppl., 157–165.

Waisman, D. M., Singh, T. J., and Wang, J. H. (1978). The modulator-dependent protein kinase: A multifunctional protein kinase activatable by the Ca^{2+}-dependent modulator protein of the cyclic nucleotide system. *J. Biol. Chem.* **253,** 3387–3390.

Walsh, M. P., Vallet, B., Autric, F., and Demaille, J. G. (1979). Purification and characterization of bovine cardiac calmodulin-dependent myosin light-chain kinase. *J. Biol. Chem.* **254,** 12136–12144.

Walsh, M. P., Guilleux, J. C., and Demaille, J. G. (1981a). Calcium- and cyclic AMP-dependent regulation of myofibrillar calmodulin-dependent myosin light chain kinases from cardiac and skeletal muscles. *Adv. Cyclic Nucleotide Res.* **14** (in press).

Walsh, M. P., Persechini, A., Hinkins, S., and Hartshorne, D. J. (1981b). Is smooth muscle myosin a substrate for the cAMP-dependent protein kinase? *FEBS Lett.* **126,** 107–110.

Watterson, J. G., Kohler, L., and Schaub, M. C. (1979). Evidence for two distinct affinities in the binding of divalent metal ions to myosin. *J. Biol. Chem.* **254,** 6470–6477.

Weiss, G. B. (1977). Calcium and contractility in vascular smooth muscle. *Adv. Gen. Cell. Pharmacol.* **2,** 71–154.

Wolf, H., and Hofmann, F. (1980). Purification of myosin light chain kinase from bovine cardiac muscle. *Proc. Natl. Acad. Sci. U.S.A.* **77,** 5852–5855.

Yazawa, M., and Yagi, K. (1978). Purification of modulator-deficient myosin light-chain kinase by modulator protein–sepharose affinity chromatography. *J. Biochem. (Tokyo)* **84,** 1259–1265.

Yazawa, M., Kuwayama, H., and Yagi, K. (1978). Modulator protein as a Ca^{2+}-dependent activator of rabbit skeletal myosin light-chain kinase. Purification and characterization. *J. Biochem. (Tokyo)* **84,** 1253–1258.

Yerna, M.-J., Dabrowska, R., Hartshorne, D. J., and Goldman, R. D. (1979). Calcium-sensitive regulation of actin–myosin interactions in baby hamster kidney (BHK-21) cells. *Proc. Natl. Acad. Sci. U.S.A.* **76,** 184–188.

6

Lipid Transport and Atherogenesis: Role of Apolipoproteins

Richard L. Jackson

I. INTRODUCTION

Although there has been a recent trend toward a decrease in mortality due to cardiovascular disease, atherosclerosis and its major clinical complication, ischemic heart disease (IHD), still remain the major cause of death in the United States today (Stern, 1979; Stamler, 1980).

VASCULAR SMOOTH MUSCLE: METABOLIC, IONIC, AND CONTRACTILE MECHANISMS

ISBN 0-12-195220-7

Epidemiological studies have provided predictive evidence that certain risk factors lead to greater occurrence of the disease. For example, elevated plasma lipids, smoking, hypertension, obesity, and diabetes have all been implicated as factors which increase the risk of developing atherosclerosis. Because of these various factors, atherosclerosis can be considered a disease of multiple interactive etiologies. A major problem in understanding the cellular mechanisms of atherogenesis in man is that the disease develops long before the clinical symptoms. Thus alterations at the cellular level occur in advance of the formation of lesions. The most striking characteristic of the advanced atherosclerotic lesion is the accumulation of large amounts of lipid as cholesteryl esters and of connective tissue elements. Proliferation of arterial smooth muscle cells is considered the major cause of the accumulation of these cellular components. What initiates this proliferation and how do the various factors account for the greater risk of developing the disease? To answer this question much attention has been given to the role of plasma lipids and lipoproteins in the development of atherogenesis. Epidemiological data, experimental studies on animals, and clinical trials have all shown that there is a relationship between plasma lipids, cholesterol, and triglycerides and atherosclerosis. A recent area of research in which there is much interest is the importance of high-density lipoproteins (HDL) in atherogenesis. It is now known that it is not the amount of total plasma cholesterol which correlates with IHD, but that which is transported in either low-density lipoproteins (LDL) or HDL. The greater the level of LDL, the greater the risk of the disease, whereas HDL have a negative correlation with risk.

The purpose of this chapter is to present new information on the role of lipoproteins in the pathogenesis of atherosclerosis. Special focus will be on the role of the apoprotein moieties. It is not the intent of this chapter to present all the historical aspects of lipoprotein structure and metabolism. Appropriate review articles will be cited for many of the primary references. Since many of the readers of this series may be less familiar with lipoproteins and atherosclerosis, it is more appropriate to present new developments and concepts with the hope of providing a perspective for future research. Many of the hypotheses on atherosclerosis are not necessarily mutually exclusive. Thus the speculation may in fact shed some new light on the cellular mechanisms.

II. LIPOPROTEIN COMPOSITION

Based on their ultracentrifugal behavior in salt solutions and electrophoretic mobility, plasma lipoproteins have been divided into four major classes (Table I) and include chylomicrons, very low-density lipoproteins (VLDL), LDL, and HDL. In addition to these major classes of lipoproteins, there are also minor lipoproteins such as Lp(a) and other lipoproteins which appear in various clinical disorders. Chylomicrons contain the greatest amount of lipid and the least amount of protein. On the other hand, HDL are much more enriched in protein. The lipoproteins which transport triglycerides are chylomicrons and VLDL, whereas LDL and HDL transport mainly cholesteryl esters and phospholipids (Table I). Although the classification system shown in Table I gives a conceptional view of lipoproteins, each of these lipoprotein classes is in fact very heterogeneous. This heterogeneity is undoubtedly related to lipoprotein metabolism and differences in lipid and protein composition. With respect to the lipid composition of lipoproteins, phosphatidylcholine and sphingomyelin are the major phospholipids. The acidic phospholipids, phosphatidylserine and phosphatidylinositol, and phosphatidylethanolamine account for only a small percentage of the total phospholipids. Cholesteryl linoleate is the major cholesteryl ester. The importance of this fact in relating lipoproteins to atherogenesis is that cholesteryl linoleate is the major lipid which accumulates in advanced atherosclerotic lesions.

The major protein (termed apoprotein) constituents of human plasma

TABLE I

Classification, Properties, and Chemical Composition of the Major Classes of Human Plasma Lipoproteins[a]

Property	Chylomicrons	VLDL	LDL	HDL
Electrophoretic mobility	Origin	Pre-beta	Beta	Alpha
Solvent density for isolation (g/ml)	1.006	1.006	1.019–1.063	1.063–1.21
Flotation rate				
S_f (1.063 g/ml)	400	20–400	0–12	—
S_f (1.21 g/ml)	—	—	—	4–9 0–4
Molecular weight	4×10^8	5.1×10^6	2.7×10^6	$1.8\text{–}3.9 \times 10^5$
Diameter (Å)	>750	250–750	200–225	40–140
Composition (wt %)				
Protein	1–2	10	21	50
Triglyceride	82	52	11	4
Cholesterol, unesterified	2	7	8	4
Cholesterol, esterified	7	12	37	14
Phospholipid	7	18	22	25

[a] From Jackson *et al.* (1976).

lipoproteins are given in Table II. The complete amino acid sequences of five of the proteins have been reported. In addition to binding and transporting lipid, some of the proteins play important roles in lipoprotein metabolism. For example, apoA-I activates lecithin:cholesterol acyltransferase (LCAT), apoC-II increases the activity of lipoprotein lipase, apoB and apoE are proteins that interact with specific binding sites on cells, and apoD has been implicated in the transfer of cholesteryl esters and triglycerides between lipoproteins. The characteristic features of each of the apoproteins and the roles they play in disease states are presented in Section IV.

III. LIPOPROTEIN STRUCTURE

In general, lipoproteins can be viewed as micellar structures with an outer monolayer of protein and polar lipids and an inner core of neutral lipids of triglycerides and cholesteryl esters (Schneider *et al.*, 1973; Bradley and Gotto, 1978). In the lipid core model, the size of the core is dependent on the amount of triglycerides and cholesteryl esters. However, the surface monolayer is the same for all classes of lipoproteins and consists of a 20-Å-thick layer of protein, phospholipid, and cholesterol.

A. Chylomicron and Very Low-Density Lipoprotein Structure

Predictions concerning the structure of chylomicrons and VLDL have been limited by the fact that these lipoproteins are very heterogeneous

TABLE II

Major Apolipoproteins of Human Plasma Lipoproteins

Apoprotein	Function
ApoA-I	Activates LCAT
ApoA-II	Unknown
ApoA-IV	Lipoprotein assembly
ApoB	Lipoprotein assembly and clearance
ApoC-I	Unknown
ApoC-II	Activates Lipoprotein Lipase
ApoC-III	Unknown
ApoD	Transfer of cholesteryl ester
ApoE	Lipoprotein clearance

and have changing protein and lipid compositions. They range in size from 2000–250 Å. Morrisett *et al.* (1977) have proposed a model for VLDL (Fig. 1). In this model cholesteryl esters are shown interspersed throughout the triglyceride core and not forming separate structures. Evidence for this type of core structure was provided by Deckelbaum *et al.* (1977b) who were unable to detect a temperature transition for cholesteryl esters as determined by differential scanning calorimetry. Thus the cholesteryl esters and triglycerides are mutually soluble in one another and do not form separate phases.

B. Low-Density Lipoprotein Structure

Human LDL from normal subjects range from 200–250 Å in diameter. Many physical techniques have been utilized to determine the struc-

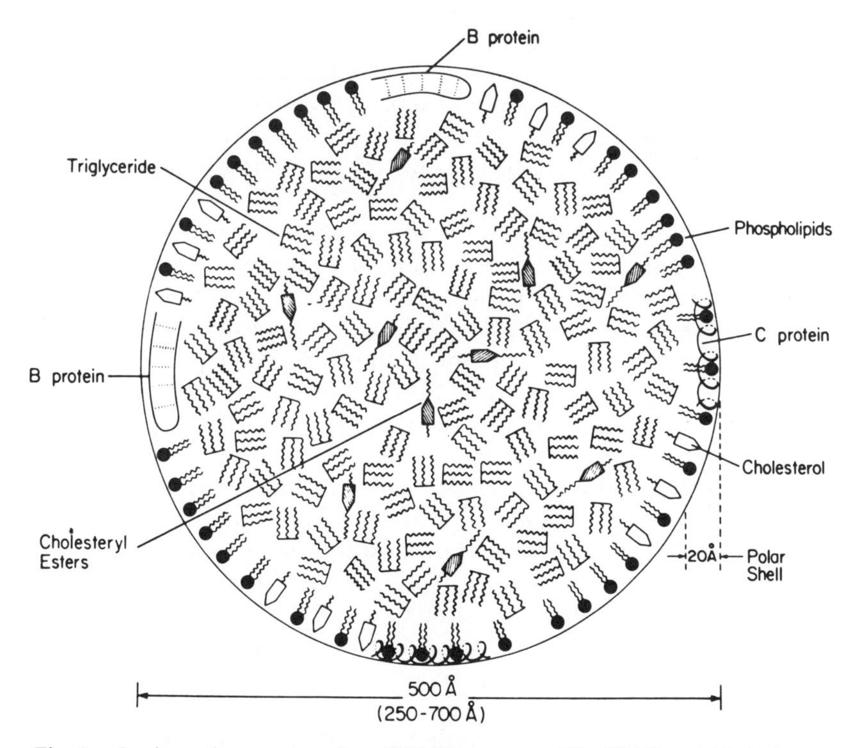

Fig. 1. A schematic representation of VLDL structure. The VLDL particle is shown as containing a central core of triglycerides and cholesteryl esters and an outer monolayer of phospholipids, unesterified cholesterol, and protein.

ture of LDL. Based on these studies, several models have been proposed (Bradley and Gotto, 1978; Kirschhausen *et al.*, 1980). The model generally accepted is that of cholesterol, protein, and phospholipid in the outer monolayer surrounding a central core of neutral lipid. Nuclear magnetic resonance studies (Yeagle *et al.*, 1977; Henderson *et al.*, 1975) show two phospholipid environments, one of which has properties similar to those of phospholipids in sonicated liposomes and one of which is immobilized and possibly associated with protein. Based on immunochemical evidence and digestion with proteolytic enzymes (Kirschhausen *et al.*, 1980), it seems certain that apoB, the major protein constituent of LDL, is at the surface of the particle. What is unclear is the distribution of the apoprotein. Based on X-ray scattering data, Luzzati *et al.* (1979) have shown that cholesterol-fed rhesus monkey LDL consist of a lipid core 250 Å in diameter, surrounded by protein which is located at tetrahedral positions on the particle. The presence of globules on the surface of LDL has also been demonstrated by freeze-etching procedures (Gulik-Krzywicki *et al.*, 1979). Cholesteryl esters, mainly linoleate, and triglycerides occupy the central core of LDL. Based on differential scanning calorimetry and small-angle X-ray scattering, Atkinson *et al.* (1977) and Deckelbaum *et al.* (1975) have shown a broad thermal transition between 20° and 45°C, which is due to a smetic-to-disordered transition of the cholesteryl esters in the core of LDL. Between these temperature ranges, the transition is reversible. At temperatures >60°C, the transition is not reversible and the lipoprotein becomes irreversibly denatured. Calorimetric studies show that there is cooperative melting of cholesteryl esters. However, their melting temperature is dramatically affected by the triglyceride content of the LDL. As the amount of triglycerides in LDL decreases from 14–1%, the melting temperature increases from 24°–35°C (Deckelbaum *et al.*, 1977a). Thus, as the cholesteryl esters become more restrained, i.e., fewer triglycerides, they become more ordered and, as a result, have a transition temperature similar to that of pure cholesteryl esters. In addition to the LDL cholesteryl ester/triglyceride ratio, the degree of saturation of the cholesteryl esters affects the transition temperature. Tall *et al.* (1977b,1978) isolated LDL from cholesterol-fed swine and monkeys. The transition temperature for both pig and monkey LDL was above body temperature. The LDL were characterized by having a large increase in cholesteryl esters and, in addition, an increase in the amount of saturated and monounsaturated cholesteryl esters. It is unclear whether the altered fatty acyl composition can account for the hypercholesterolemia present in these atherogenic-fed animals. However, as suggested by Tall *et al.* (1978), altered cholesteryl ester composition may affect the lysosomal acid lipase degradation of the more ordered cholesteryl esters.

C. High-Density Lipoprotein Structure

The structure of HDL has been studied extensively by a variety of physicochemical techniques. From these data, a number of models have been proposed (Bradley and Gotto, 1978). Based on all the known information concerning HDL structure, Edelstein *et al.* (1979) have constructed a space-filling model for HDL_3 with a chemical composition of 51 phospholipids, 13 unesterified cholesterols, 32 cholesteryl esters, and 9 triglycerides per HDL_3 particle. In this HDL model, the dimensions of all the components were derived from space-filling atomic models. The model was constructed with several basic premises. First, all the neutral lipids, cholesteryl esters, and triglycerides occupy the core of the particle. Second, all the phospholipids and cholesterol are at the surface, the phospholipids being randomly distributed such that there is no clustering. Third, the protein is placed between the polar head groups of the phospholipids, thereby forming a tightly packed surface and shielding the hydrophobic core from the aqueous environment.

The experimental evidence for a HDL core containing neutral lipids is based on low-angle X-ray scattering data. Shipley *et al.* (1972) and Laggner *et al.* (1973) determined the electron density pattern for HDL and reported that they had a central core with an electron density of 0.312 $e/Å^3$, a typical value for neutral lipids. The radius of the inner core is well defined, whereas it is more difficult to define clearly the outer monolayer. Although low-angle X-ray scattering techniques give some estimate of the size and electron density of HDL, this technique does not provide a detailed understanding of the molecular details involving each of the constituents. For this reason, HDL have also been studied by differential scanning calorimetry, fluorescence methods, and high-field nuclear magnetic resonance. Tall *et al.* (1977c) were unable to detect a cholesteryl ester thermal transition in HDL between 0° and 69°C. At temperatures >70°C, there was a broad endothermic transition, but this was due to the irreversible release of apoA-I. In contrast to those of intact HDL, the extracted lipids of HDL show a well-defined thermal transition which coincides with that of the cholesteryl esters. Because of the limiting amount of cholesteryl esters in HDL, Tall *et al.* (1977c) conclude that the lipid core domain is too small and prevents the formation of organized lipid structures, thus there is no cooperative melting of the lipids. Calorimetric data also suggest that the core is highly viscous.

Jonas (1977) has utilized 1,6-diphenylhexatriene to probe the structure of LDL and HDL and has found that the core structure of LDL is relatively more fluid than the core of HDL. It is also more fluid than isolated HDL lipids. Because of the constraints on the molecular packing of the cholesteryl esters in the limited core volume of HDL_3 it is impossible to construct the model so that the fatty acyl groups of the cholesteryl esters are extended and parallel as in LDL. In HDL, the cholesteryl esters are bent so that they are confined to the inner core. Hamilton and Cordes (1978) have used high-field natural abundance nuclear magnetic resonance and have been unable to show molecular interactions between phospholipid fatty acyl chains and cholesteryl esters.

Experimental evidence for phospholipids located in the outer monolayer of HDL and exposed to the aqueous phase is provided by enzymic studies with phospholipase A_2. Pattnaik *et al.* (1976) reported that the enzyme quantitatively hydrolyzed phosphatidylcholine and phosphatidylethanolamine in HDL. In addition, the kinetics of release indicate that all the phospholipids are equally available for hydrolysis.

The location of free cholesterol positioned between the fatty acyl chains of phospholipids and in the neutral lipid core is based on nuclear magnetic resonance. Avila *et al.* (1978) have shown that there is interaction between the steroid nucleus of cholesterol and the fatty acyl chains of the phospholipids. The exact location of the hydroxyl group of cholesterol is unknown. In the model of Edelstein *et al.* (1979) the hydroxyl group is protected from the aqueous phase by the protein moiety. However, it does seem likely that the hydroxyl group would be in contact with the water, since it must be available for esterification by lecithin:cholesterol acyltransferase.

The organization of the protein moieties of HDL has received the most attention, since they play important roles in binding and transporting lipid and modulating lipoprotein metabolism. The HDL are heterogeneous and can be divided into at least three subfractions (Anderson *et al.,* 1978). The protein composition of each subfraction is undoubtedly different, although detailed chemical analysis has not been described. In general, the two major proteins of human HDL are apoA-I and -II. These two proteins constitute about 90% of the total HDL protein, with an apoA-I/apoA-II weight ratio of approximately 3:1 to 2:1. The remainder of the HDL proteins consist of apoC's, apoE, and possibly apoD. Several generalizations can be made concerning the properties of apoA-I and -II, which form the basis for preferentially placing the proteins at the surface of HDL and include the following: Apoproteins form stable monolayers; they form amphipathic helices when associated with phospholipid, and they are accessible to antibodies and cross-linking reagents. Jackson *et al.* (1979) and Shen and Scanu (1980) have shown that apoproteins are surface-active and form stable monolayers in the absence of lipid. Apoproteins also interact with phospholipids at an interface and, in the case of apoC-II and -III, remove phosphatidylcholine from the interface (Jackson *et al.*, 1979). The surface-active properties of apoA-I and -II can be explained by an examination of space-filling models of their structures. A common feature of all the plasma apolipoproteins is that they are highly helical in the presence of lipid. Based on an inspection of space-filling models, Segrest *et al.* (1974) have reported that the helices have two faces, one hydrophobic and one hydrophilic. The apolar face is in contact with the hydrophobic lipids, whereas the polar face is associated with the polar group of the phospholipid and the water phase. Thus the thermodynamic stability of apoproteins at the surface of the lipoprotein particle involves their ability to assume these helical structures (Segrest, 1977; Massey *et al.*, 1979). The surface orientation of apoproteins is also consistent with the finding that apoA-I and -II are accessible to the bifunctional cross-linking reagent 1,5-difluoro-2,4-dinitrobenzene. Grow and Fried (1978) have reported that apoA-I and -II form products with each other but not with themselves after incubation with this cross-linking agent.

IV. APOPROTEIN STRUCTURE, METABOLISM, FUNCTION, AND ROLE IN DISEASE

Much of the detailed information concerning the chemical and physical properties of each of the individual apoproteins has been described in several recent reviews (Jackson *et al.*, 1976; Steinberg, 1979; Schaefer *et al.*, 1978b; Smith *et al.*, 1978; Osborne and Brewer, 1977). In the present discussion, we have limited our review to only recent information concerning these proteins and, in particular, their relationship to lipoprotein metabolism and atherosclerosis.

A. ApoA

1. ApoA Structure

ApoA-I and -II are the major protein constituents of HDL. The amino acid sequences of both proteins are known (Baker *et al.*, 1974, Brewer *et al.*, 1972, 1978). In the absence of lipid, both apoA-I and -II self-associate in aqueous solutions, forming higher-molecular-weight aggregates (Osborne and Brewer, 1977). The degree of self-association is dependent on the protein concentration. At protein concentrations >0.1 mg/ml there are increased numbers of secondary structures. The forces which stabilize the protein are primarily hydrophobic in nature and are undoubtedly due to the hydrophic face of the amphipathic helix. Tall *et al.* (1976) have measured the enthalpy of unfolding of apoA-I by differential scanning caloriometry; apoA-I unfolds between 43° and 70°C with an enthalpy of unfolding of 64 kcal/mol. Massey *et al.* (1979) have measured the energetics of association of apoA-II with phospholipid by microcalorimetry and reported a value for the enthalpy of association of 52 kcal/mol. There was a direct correlation of the enthalpy of association with the increase in helicity, suggesting that helix formation is the driving force for lipid–apoprotein interaction. ApoA-I solubilizes dimyristoyl phosphatidylcholine, forming structures which appear as disks when examined by electron microscopy (Tall *et al.*, 1977a; Wlodawer *et al.*, 1979). The composition of the isolated lipid–protein complexes is dependent on the initial weight ratio of the protein and lipid. As discussed below, discoidal structures resembling those of apoA-I and phospholipid have been observed in the perfused rat and during *in vitro* catabolism of chylomicrons and may represent nascent HDL.

In man and baboon, apoA-II is a protein containing two identical chains linked by a disulfide at residue 6. However, in other animals the disulfide bond is absent, and the protein is present in a monomeric form.

2. *ApoA Synthesis*

Most of the information concerning apoA-I synthesis has been derived from studies on the rat liver (Hamilton *et al.*, 1976). Very little is known about apoA synthesis in man. In the rat, and under conditions in which LCAT is inhibited, the liver produces HDL particles which appear discoidal in structure and enriched in apoE (Section IV,E), the ratio of apoE to apoA-I being 10:1. In the absence of LCAT, the ratio is 1:7 (Felker *et al.*, 1977). The intestine has also been shown to produce nascent discoidal HDL (Green *et al.*, 1978). ApoA-I has been identified in human intestinal tissue by quantitative immunochemical methods (Schonfeld *et al.*, 1978; Glickman and Green, 1977). The amount present in the intestine may quantitatively account for all the apoA-I in the plasma compartment. During fat absorption, there is a marked increase in intestinal synthesis of apoA-I. After injecting ^{125}I-labeled chylomicrons into postheparin plasma, Schaefer *et al.* (1978c) found that almost all the labeled apoproteins were recovered in HDL. These results suggest that apoA proteins enter the plasma compartment by either direct synthesis in the gut or liver or as part of chylomicrons, and that during catabolism HDL are produced. Tall and Small (1978) have presented a model which attempts to explain how HDL are produced from catabolism of triglyceride-rich lipoproteins. The basis for this model is that during hydrolysis of core triglycerides there is an excess of the surface components, phospholipids and proteins, which form projections. These projections then "pinch off" and form discoidal HDL which become spherical by the action of LCAT.

3. ApoA Catabolism

Studies on the catabolism of HDL are complicated by the fact that apoproteins are freely exchangeable between lipoproteins and that HDL are very heterogeneous. Blum *et al.* (1977) have measured the rate of disappearance of ^{125}I-labeled HDL and found that the rates of catabolism of apoA-I and -II are identical. The tissue site of apoA degradation is not firmly established in man. In animals, the liver has been implicated (Quarfordt *et al.*, 1980; Van Berkel *et al.*, 1980). In man, subjects with nephrotic syndrome, diabetes, or hyperchylomicronemia all show very low levels of HDL. Patients with renal disease also have an increased amount of apoA-I in the urine.

As discussed above, there are at least three major components of HDL, including HDL_{2a}, HDL_{2b}, and HDL_3 (Anderson *et al.*, 1978). The metabolic interrelationship between HDL_2 and HDL_3 is dependent on triglyceride catabolism. Patsch *et al.* (1978) have used an *in vitro* system of human VLDL, HDL_3, and bovine milk lipoprotein lipase and have shown that HDL_3 is quantitatively converted to an HDL_2-like particle. Forte *et al.* (1979) have confirmed the production of HDL_2 from HDL_3 by measuring the increase in HDL_2 after heparin injection *in vivo*. However, in the studies of Forte *et al.* (1979), HDL_3 did not disappear but remained at the same preheparin concentration. The mechanism by which the surface components of VLDL transfer to HDL_3 are unknown. The lipids and proteins could transfer as individual components. It is also possible that the surface components form a discoidal structure during lipolysis and that this structure fuses with HDL_3 to yield HDL_2.

4. ApoA Function

The major metabolic role of apoA-I is in the activation of LCAT (Fielding *et al.*, 1972). This enzyme is of hepatic and intestinal origin and circulates in the plasma compartment. It has been purified to hemogeneity (Aron *et al.*, 1978; Chung *et al.*, 1979; Albers *et al.*, 1976). The enzyme is a glycoprotein with an apparent molecular weight of approximately 60,000–70,000. It is multifunctional and must first hydrolyze a C_2 fatty acid from phospholipid and then transfer it to cholesterol. Aron *et al.* (1978) have demonstrated that the phospholipase A_2

activity of LCAT is dependent on the amount of cholesterol in the substrate. With low concentrations of cholesterol, the phospholipase activity is decreased as compared to that with high levels of cholesterol. For maximal activation of LCAT by apoA-I, Chung *et al.* (1979) have reported that there must be six apoA-I molecules per phospholipid particle; the addition of apoA-II to the apoA-I–phospholipid complex causes a reduction in activity by dissociating the apoA-I from the lipid. Recently Yokoyama *et al.* (1980) prepared a synthetic docosapeptide which did not correspond to an amino acid sequence in apoA-I but possessed an amphipathic helix and activated LCAT. The maximum reaction rates for the peptide were 18% of that of apoA-I for cholesterol esterification and 50% for release of fatty acids from phospholipid vesicles. The results with the synthetic peptide suggest that apoA-I modulates the structure of the lipid, which in turn increases the activity. Pownall *et al.* (1980) have also synthesized a model lipid-associating peptide of 20 amino acid residues and reported an activity of 65% of that of apoA-I.

There have been no reports of a metabolic role for apoA-II. In fact, the amount of apoA-II in HDL from the dog and pig is exceedingly low or absent.

5. *ApoA-I and Disease*

The plasma levels of apoA-I and -II have been reported by many groups employing several different methods. In normal human subjects, the amount of apoA-I in plasma is 120 mg/dl, and of apoA-II, 40 mg/dl. The levels of apoA-I and -II are higher in women than in men. Blum *et al.* (1977) determined that the rate of catabolism of apoA-I was the same in men and women. As discussed below, this difference in HDL levels may be related to differences in the rate of lipoprotein triglyceride metabolism.

To date there have been no patients described who show an absence of apoA proteins. However, there are several clinical situations where there are decreased levels of HDL and, consequently, apoA proteins. In Tangier disease there are very low concentrations of plasma apoA-I and -II (Schaefer *et al.,* 1980). Glickman *et al.* (1978) and Schaefer *et al.* (1978a) have reported that the defect in Tangier disease is not due to a lack of apoA-I synthesis but that the catabolism of apoA proteins is very rapid when compared to that in normals. Because of the rapid turnover of apoA, the level of HDL is low, and thus there is no mechanism for the transport of free cholesterol. As a result, the free cholesterol is taken up by macrophages, and esterified to cholesteryl oleate. A characteristic feature of Tangier patients is the accumulation of cholesteryl oleate in tissues. Thus, because of the lack of a suitable acceptor for cell membrane cholesterol, the sterol is stored in the tissue as the ester.

B. ApoB

1. ApoB Structure

ApoB is the major protein constituent of chylomicrons, VLDL, and LDL. Limited information is available concerning the properties of apoB. The apoprotein is extremely difficult to solubilize and to dissociate. Reported molecular weights range from 8000 to 275,000 (Bradley *et al.,* 1978; Socorro and Camejo, 1979; Deutsch *et al.,* 1978; Steele and Reynolds, 1979). Based on a limited number of cyanogen bromide peptides which can be identified, Deutsch *et al.* (1978) and Bradley *et al.* (1978) have suggested that the apoprotein has a molecular weight in the range of 25,000–30,000.

2. ApoB Metabolism

ApoB is synthesized in both the intestine and liver. Its synthesis is inextricably related to the production of triglycerides. In the rare autosomal recessive disorder in man known as abetalipoproteinemia, apoB is not synthesized, and consequently chylomicrons, VLDL, and LDL are completely absent in the plasma (Glickman *et al.,* 1979). In these rare subjects there is an accumulation of lipid in the intestine, since dietary fat which is absorbed cannot be assembled into chylomicrons.

Very little detailed information is known concerning the assembly, secretion, and metabolic control involved in lipoprotein production. It can be assumed that it must involve synthesis of the protein and lipid constituents, assembly of these various components, and finally release of the lipoproteins into the lacteals and bloodstream. ApoB is presumably synthesized on membrane-bound polyribosomes. It is then released into cisternae of the endoplasmic reticulum. Although it may not be directly applicable in man, Chan *et al.* (1978) have isolated in mRNA for a VLDL protein from the liver of estrogen-stimulated chickens. The administration of estrogen causes an increase in hepatic production of VLDL. The apoprotein produced has been isolated, and its amino acid sequence has been determined. Chan *et al.* (1978) have translated the purified mRNA for the apoprotein in a cell-free system. The product of translation has been sequenced; the mRNA produces a protein which contains a 23-amino-acid-residue leader sequence. Leader or signal sequences have also been identified for chicken apoA-I and albumin. Thus it appears that apoproteins are similar to other secretory proteins in that they are synthesized as larger precursors and that the special signal sequence is removed after it serves its function in initiating transfer across the endoplasmic membrane.

Triglycerides and phospholipids are synthesized in the smooth endoplasmic reticulum (SER). Phosphaditylcholine is synthesized on the outside of the SER; the lipid presumably must then undergo transmembrane migration to the inside of the bilayer. A central unanswered problem involves the mechanisms by which the newly synthesized protein acquires the lipid moieties. After packaging together of the lipid and protein, the nascent VLDL passes through the Golgi apparatus where carbohydrate is attached to the protein moieties. The VLDL particles isolated from the Golgi apparatus have a composition very similar to that of plasma VLDL, with the exception of a lack of apoC proteins. The final step in chylomicron or VLDL synthesis involves secretion. Since colchicine and vincristine block the secretion of VLDL, it can be assumed that the microtubular transport system is involved in the release of VLDL from hepatocytes. Orotic acid, cyclohexamide, and ethionine also inhibit the release of VLDL, presumably by causing aggregation of the contractile protein system.

Schaefer *et al.* (1978b) have recently reviewed the pertinent clinical data concerning the kinetics of apoB synthesis. In normal subjects, the amount of apoB synthesized in the form of VLDL is about 10 mg kg^{-1} day^{-1}. Most of the apoB which appears in nascent VLDL is converted to LDL by a unidirectional delipidation pathway which involves lipoprotein lipase. Synthesis of LDL apoB ranges between 12–14 mg kg^{-1} day^{-1}, values which are in agreement for VLDL apoB synthesis. The contribution of the apoB in chylomicrons to the plasma LDL apoB pool is unknown. It is of considerable interest, however, that Kane *et al.* (1980) have recently shown that the apoB in chylomicrons is chemically different from the apoB in VLDL and LDL. Because of the uncertainty of the role of chylomicron apoB in LDL synthesis, we will limit our discussion to only the catabolism of VLDL.

The evidence for the VLDL delipidation pathway was derived from turnover studies using VLDL labeled with ^{125}I in the protein moieties. Since apoB does not exchange between lipoprotein particles, it was possible to follow the kinetics of decay of VLDL and to show that the final products were LDL. The key enzyme and apoprotein in the hydrolysis of VLDL triglycerides is lipoprotein lipase and apoC-II (Fig. 2); apoC-II is a specific activator protein for this enzyme. When chylomicrons and VLDL enter the circulation, they lack apoC proteins. The first metabolic alteration to occur upon entering the circulation is the transfer of apoC

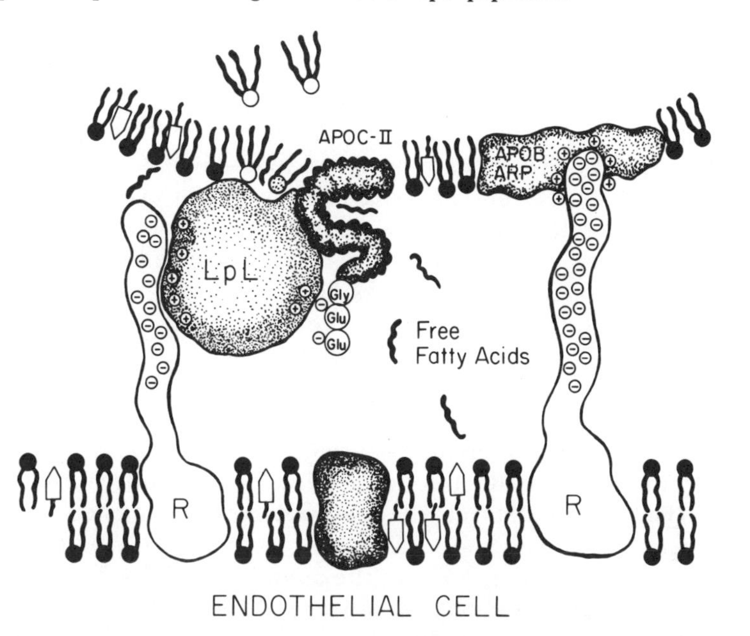

Fig. 2. A schematic representation of the interaction of lipoprotein lipase (LpL) with glycosaminoglycans at the endothelial surface. The LpL is shown as interacting with specific glycosaminoglycan receptors (R) by ionic interactions. ApoC-II is shown at the surface of a VLDL particle, interacting with a specific region of LpL. ApoB and the arginine-rich protein (ARP or apoE) are shown interacting with specific cell surface receptors (R).

from HDL to the surface of VLDL and chylomicrons; the distribution of apoC-II appears to be dependent only on the relative concentration, i.e., surface area, of the lipoproteins (Kashyap *et al.,* 1978). Havel *et al.* (1973) have shown that after an oral fat load there is an immediate transfer of apoC to chylomicrons and that during lipolysis the apoC returns to HDL, in particular to HDL_2. The hydrolysis of chylomicrons and VLDL triglycerides is mediated by lipoprotein lipase. The importance of the enzyme in lipoprotein metabolism is exemplified by its absence in a rare autosomal recessive disorder termed type I hyperlipoproteinemia. In these subjects there is an absence of adipose tissue lipoprotein lipase (Krauss *et al.,* 1974) and, as a result, patients have exceedingly high levels of chylomicrons. The characteristic features of lipoprotein lipase are that it is released into the circulation after heparin injection, it is inhibited by high salt concentrations, and it is activated by apoC-II. The importance of apoC-II in lipoprotein metabolism is exemplified by the fact that the rare individuals who have a genetic deficiency of the activator peptide have marked hypertriglyceridemia (Cox *et al.*, 1978).

Since lipoprotein lipase is released from the luminal surface of the endothelial cell by heparin, it has been suggested that the enzyme is bound to specific cell membrane glycosaminoglycans. In this regard, Bengtsson and Olivecrona (1977) have reported that a variety of sulfated polysaccharides release lipoprotein lipase. The mechanism by which the lipoproteins interact with lipoprotein lipase is controversial. However, most evidence suggests that lipoprotein triglycerides are hydrolyzed at the endothelial surface (Fig. 2). On the other hand, Felts *et al.* (1975) have proposed that, as a result of interaction of the lipoprotein with the endothelial cell, the enzyme is released and becomes part of a lipoprotein–lipoprotein lipase complex. Regardless of the mechanism, the kinetic properties of endothelial cell-bound lipoprotein lipase in the perfused rat heart are identical to those of the heparin-released soluble enzyme (Fielding and Havel, 1977). An interesting consistent finding is that larger chylomicron and VLDL particles are hydrolyzed faster than smaller particles. Furthermore, as the amount of triglyceride decreases to about 25% during catabolism, the rate of catabolism decreases. The decrease in lipolysis may be due to a decreased catalytic constant (K_{cat}) and/or to a change in the K_m of the enzyme for the substrate. It is also possible that the decrease in activity is due to a loss of apoC-II or a change in the surface properties of the VLDL. For example, it is known that during lipolysis there is an increase in the relative amounts of cholesterol, sphingomyelin, and apoB. This alteration in surface monolayer composition undoubtedly affects the surface pressure of the resultant lipoproteins and may affect further catabolism. Another possibility explaining the decrease in the rate of lipolysis is that free fatty acids may affect the interaction of lipoprotein lipase with the substrate or with apoC-II. Scow *et al.* (1979) have used monolayer techniques to demonstrate that with limiting amounts of albumin there is an accumulation of free fatty acids at the interface and a decrease in lipolysis. This finding is also of interest in that it relates to the transfer of the products of lipolysis from the plasma compartment to the endothelial cell and, as in the case of adipocytes, resynthesis and storage of triglycerides. Based primarily on electron microscopic evidence, Scow *et al.* (1980) have proposed that the surface film of the chylomicron and the plasma membrane of the endothelial cell fuse together. As a result, there is lateral movement of monoglycerides and free fatty acids on the lipoprotein surface and in the cell membrane.

In addition to the loss of triglycerides during lipolysis, phospholipid, mainly phosphatidylcholine, unesterified cholesterol, and apoC, become depleted. The loss of phospholipid occurs primarily as a result of the phospholipase activity of lipoprotein lipase. It is possible that plasma phospholipid exchange proteins may also be involved in the transfer of phospholipids between lipoproteins (Section IV,D).

3. *ApoB Function*

From the above discussion, it can be concluded that apoB is synthesized as part of a VLDL particle which functions to deliver triglycerides from the liver to either muscle for energy utilization or to adipose tissue for storage. As a result of the action of lipoprotein lipase, a triglyceride-poor particle or remnant is produced, which is converted to LDL. The mechanism for the conversion of the remnant to LDL is unknown. The liver may be involved, but direct proof is still lacking. Regardless of the mechanism, the final product of VLDL catabolism is LDL, a lipoprotein particle which contains only apoB and is enriched in cholesteryl esters. The catabolism of apoB, i.e., as LDL, occurs primarily by a receptor-mediated process in peripheral tissues (for review, see Goldstein *et al.*, 1979; Brown and Goldstein, 1979). The function of apoB is to serve as a recognition protein on the surface of the lipoprotein in order to recognize the LDL receptor on the cell membrane. With the exception of the nervous system, all nonhepatic tissues so far examined have been shown to possess specific receptor molecules for LDL. Ultrastructural studies have shown that LDL receptors are not randomly distributed on cells but are clustered in regions called coated pits. The chain of events culminating in the internalization of LDL by endocytosis and delivery of cholesterol to the cell is illustrated in Fig. 3. The salient points in the LDL pathway are that during catabolism apoB is degraded to amino acids, and cholesteryl esters to free cholesterol. The lipid taken up by this process has three effects on cholesterol metabolism. By still unknown mechanisms, there is suppression of LDL receptor synthesis, activation of microsomal fatty acyl-CoA cholesterol acyltransferase (ACAT), and inhibition of 3-hydroxy-3-methylglutaryl coenzyme A (HMG-CoA) reductase. In order to remove free cholesterol from the cell, it must first be transported to the cell membrane where it is removed by transfer to an acceptor molecule, presumably HDL. In the absence of an appropriate acceptor, the cholesterol is reesterified by ACAT and the cholesteryl esters are stored in the cell. The role of HDL in the cellular mechanisms of atherosclerosis is discussed in more detail in Section V,C.

4. *The Role of ApoB in Disease*

As discussed above, the absence of apoB, as in the case of abetalipoproteinemia, is associated with a near lack of chylomicrons, VLDL, and

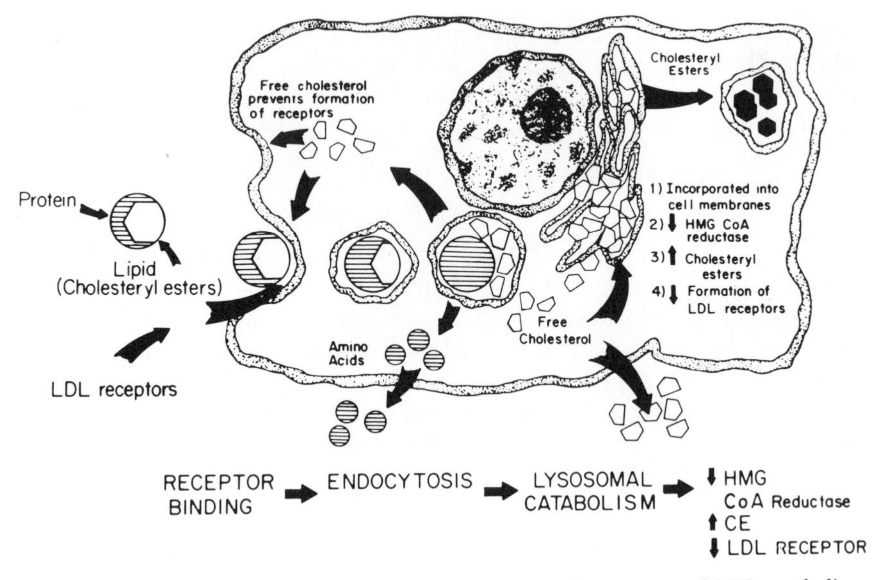

Fig. 3. Metabolism of LDL by extrahepatic tissues. The process of LDL catabolism includes binding of the lipoprotein to specific LDL receptors, endocytosis of the lipoprotein, breakdown of LDL and cholesteryl esters (CE) by lysosomal enzymes, and the effects of unesterified cholesterol on HMG-CoA reductase, acyl-CoA cholesterol acyltransferase, and synthesis of LDL receptors. For further descriptions of this pathway refer to Goldstein and Brown (1977).

LDL in the circulation. The major problem concerning LDL catabolism is the lack of a specific receptor molecule which recognizes the apoB in LDL. In homozygotes who lack the LDL receptor, plasma cholesterol levels exceed 800 mg/dl and the risk of developing atherosclerosis is greatly accelerated.

C. ApoC

1. ApoC Structure

The amino acid sequences of apoC-I, -II, and -III are known and consist of 57, 78, and 79 residues, respectively (Smith *et al.,* 1978). Each of the apoproteins avidly associates with phospholipids. The mechanism for lipid association involves the interaction of amphipathic helixes in the apoproteins with the phospholipids.

2. ApoC Metabolism

Limited information is available concerning the tissue responsible for the synthesis of apoC in man (Schonfeld *et al.,* 1980). In the rat, apoC proteins are secreted from the perfused liver. Whether the apoC proteins are released from the liver alone or bound to HDL or VLDL is not known. In fasting subjects, most of the apoC-II and -III is associated with VLDL and HDL. Kashyap *et al.* (1977) have measured the amount of apoC-II in plasma of normal subjects and subjects with hypertriglyceridemia. The level of apoC-II in normal subjects ranges between 4-5 mg/dl, with no apparent difference in males or females. The amount of apoC-II in subjects with hypertriglyceridemia is increased over that in normal subjects. However, when the amount of apoC-II per unit of surface area is calculated, there is an absolute decrease in the apoC-II content per lipoprotein particle.

Berman *et al.* (1978) have analyzed the apoC turnover data in man and conclude that apoC can be modeled in four compartments in VLDL and in two HDL compartments. Based on this model 100-400 mg of apoC-II is synthesized per day, with an apoC plasma half-life ranging from 10-18 hr. As described above, apoC proteins are in equilibrium between VLDL and HDL; during lipolysis they recycle to HDL. The site of degradation of apoC proteins is unknown. It is of interest that the half-life of apoC is considerably shorter than the 5 days for apoA-I. It may well be that apoC proteins are catabolized as a unit separate from HDL.

3. *ApoC Function*

The one apoC protein which clearly plays an important role in lipoprotein metabolism is apoC-II (Fig. 2). Lipoprotein triglycerides or long-chain artificial triglyceride substrates are not hydrolyzed by lipoprotein lipase in the absence of apoC-II. Kinnunen *et al.* (1977) have attempted to identify the sequence requirement within apoC-II which is necessary for activation of the enzyme. Based on studies with cyanogen bromide peptides and various synthetic peptides, it has been concluded that the minimal sequence requirement in apoC-II resides between residues 55 and 78. Removal of the three carboxyl-terminal residues of Gly-Glu-Glu abolishes activity. Since it is well known that high anionic salts inhibit lipoprotein lipase activity, it has been suggested that there is ionic interaction between the carboxylate ions and positively charged portions of the enzyme (Kinnunen *et al.,* 1977).

Monolayer techniques have been utilized to study the molecular details of the interaction of apoC-II and lipoprotein lipase. Miller and Smith (1973) used partially purified bovine milk lipoprotein lipase and demonstrated that apoC-II interacted with the enzyme even in the absence of substrate. In a more recent report, Jackson *et al.* (1980) have shown that the enzymic activity of lipoprotein lipase is highly dependent on surface pressure. At low surface pressure, the enzyme is irreversibly inactivated at the lipid interface. At a surface pressure of 10 dynes/cm, apoC-II protects the enzyme from surface denaturation. This protection is also observed with other apoproteins. However, at 25 dynes/cm, only apoC-II shows a specific activation. With these controlled conditions for maintaining constant surface pressure, the enzyme is maximally activated at a molar ratio of apoC-II to lipoprotein lipase of 200:1, indicating a very low association constant.

4. The Role of ApoC in Disease

Plasma apoC concentrations are affected by a variety of dietary, hormonal, and metabolic parameters. Falko *et al.* (1980) have shown that a high carbohydrate diet decreases the apoC content compared to a high fat diet. A major unanswered question is whether apoC-II modulates the rate of triglyceride clearance in man. Recently a patient has been identified who lacks apoC-II (Cox *et al.*, 1978) and, as a result, has hypertriglyceridemia. After a transfusion with normal plasma, the plasma triglycerides are normalized. Catapano *et al.* (1979) have isolated the VLDL from a patient with apoC-II deficiency and have demonstrated in an *in vitro* experiment with purified bovine milk lipoprotein lipase that the rate of triglyceride hydrolysis is dependent upon the addition of apoC-II. The apoC-II-deficient subject clearly demonstrates the importance of the apoprotein in lipoprotein metabolism. However, other evidence shows that it does not play a regulatory role under normal circumstances. For example, patients who are heterozygous for the apoC-II deficiency and have apoC-II levels one-half of normal have normal plasma triglycerides. In addition, Fielding and Fielding (1976) reported that rat lymphatic chylomicrons containing only one-tenth the amount of apoC-II as plasma chylomicrons were hydrolyzed at the same rate in a perfused rat heart system. These studies suggest that apoC-II does not play an important regulatory role under ordinary conditions. What appears to play an important regulatory role is the amount of lipoprotein lipase itself, which is discussed below in relation to the concentration of plasma HDL.

D. ApoD

1. ApoD Structure

ApoD was first isolated as part of a subclass of HDL_3. McConathy and Alaupovic (1976) subjected HDL_3 to chromatography on hydroxyapatite and concanavalin A-Sepharose and identified a distinct liproprotein class which was different from the apoA class of HDL_3. The isolated complex of McConathy and Alaupovic contained 70% protein and 30% lipid. The major protein of this class of lipoprotein had a molecular weight of 22,100. The protein was characterized as being a glycoprotein with half-cysteine present. Chajek and Fielding (1978) have also purified a protein from HDL_3 by immunoaffinity chromatography with anti-apoD attached to Sepharose. The amino acid composition of the isolated protein is similar to that of McConathy and Alaupovic (1976). However, Chajek and Fielding (1978) report a molecular weight of 35,000.

2. *ApoD Metabolism*

Nothing is known concerning the site of synthesis or catabolism of apoD. As discussed below, a possible function of apoD is to transfer cholesteryl esters and triglycerides between lipoprotein particles. In this regard, Ihm *et al.* (1980) have recently shown that the rat is deficient in cholesteryl ester transfer protein. Clearly much needs to be learned about the metabolism of apoD.

3. *ApoD Function*

Nichols and Smith (1965) were the first to report that cholesteryl esters transferred between lipoproteins. However, the amount of transfer was relatively low compared to lipoprotein metabolism. Consequently this early finding was ignored until Zilversmit *et al.* (1975) isolated a protein from human plasma, which facilitated the exchange of cholesteryl esters between LDL and VLDL. The protein fraction was isolated (Pattnaik *et al.*, 1978) by chromatography of lipoprotein-free plasma on phenyl-Sepharose followed by DEAE-cellulose chromatography and isoelectric focusing. Chajek and Fielding (1978) have also isolated a cholesteryl ester transfer protein from human HDL by immunoaffinity chromatography on a column of anti-D-Sepharose. The properties of this protein are clearly different from those reported by Pattnaik *et al.* (1978). The preparation, as described by Chajek and Fielding (1978), transfers cholesteryl esters from HDL to VLDL, and then there is a transfer of VLDL triglycerides to HDL. The physiological significance of cholesteryl ester and triglyceride transfer remains to be determined. Based on the low specific activities it appears that the transfer protein is not important in metabolism. However, it must be considered that HDL and LDL have half-lives on the order of 2–4 days. In this regard, Barter and Lally (1979) concluded that all the cholesteryl esters in plasma VLDL could be derived from HDL. It is not known in what phase of VLDL catabolism the cholesteryl esters are transferred from HDL to VLDL.

E. ApoE

1. ApoE Structure

Shore and Shore (1973) were the first to isolate a protein from human plasma which was rich in arginine. Based on its high content of arginine, this protein was initially called arginine-rich protein. Subsequently, it was termed apoE. The protein is present in several different polymorphic forms (Utermann *et al.,* 1977; Zannis and Breslow, 1980). The heterogeneity is presumably due to differences in carbohydrate content, although a detailed structural analysis has not been performed. ApoE can be purified from lipid-free VLDL by affinity chromatography on heparin–Sepharose (Shelburne and Quarfordt, 1977). In man, the protein has a molecular weight of 33,000–39,000 (Shelburne and Quarfordt, 1974; Weisgraber *et al.,* 1980). Weisgraber and Mahley (1978) have also isolated a form of the protein in low yield, which has a higher molecular weight. This form of apoE consists of a mixed disulfide of apoE and apoA-II. The significance of this abnormal form of apoE is unknown.

2. ApoE Metabolism

The sites of synthesis of apoE in man are not completely known. In the rat, apoE enters the circulation with nascent HDL particles. Hamilton *et al.* (1976) reported that, in the presence of an LCAT inhibitor, apoE was the major HDL apoprotein in the perfused rat liver. In the absence of the inhibitor, the major apoprotein was apoA-I, suggesting that during the esterification of cholesterol by LCAT, apoE transfers from HDL to VLDL. In man, apoE is mainly found in VLDL. Diets rich in cholesterol are associated with an increase in the plasma levels of apoE in man and in other species. ApoE accumulates in a VLDL-like lipoprotein particle which migrates with beta mobility. In addition, Pitas *et al.* (1979) have isolated an apoE-containing lipoprotein (termed HDL_c) which floats with HDL. It binds to the cholesterol transport receptors (Fig. 3) about 10–100 times more readily than lipoproteins containing only apoB. The apoE-containing lipoprotein also binds with a higher affinity ($K_d = 0.12 \times 10^{-9} M$) as compared to dog LDL ($K_d = 2.8 \times 10^{-9} M$). These findings suggest that apoE is catabolized by peripheral tissue and serves as a determinant on the lipoprotein particle in recognizing specific cell surface receptors.

3. ApoE Function

A receptor role for apoE was also demonstrated by Sherrill *et al.* (1980) in the liver for the uptake of chylomicron remnants. These investigators reported that the perfused rat liver displayed a high-affinity, receptor-mediated uptake and saturation kinetics for dog HDL_c. The administration of pharmacological doses of estradiol to rats markedly increased the hepatic uptake of apoE-containing lipoproteins. Chao *et al.* (1979) treated rats with large amounts of ethinyl estradiol, causing marked hypolipidemia. The rate of removal of apoE-containing lipoproteins is greatly enhanced in perfused livers from estrogen-treated rats. The mechanisms by which estrogen increases the catabolism of these lipoproteins are not entirely clear. Increased receptor activity may reflect an estrogen-induced increase in the expression of receptors or in the affinity of the receptor lipoprotein interaction, or it may also alter other apoproteins on the surface of apoE-containing lipoproteins. In this regard Windler *et al.* (1980a,b) have reported that one or more apoC proteins might oppose the effects observed for apoE in the hepatic uptake of triglyceride-rich lipoproteins in rat liver.

4. The Role of ApoE in Disease

Type III hyperlipoproteinemia (familial dysbetalipoproteinemia) is a monogenic human disorder of lipoprotein metabolism. The absence of one of the isoforms of apoE may explain the accumulation of β-VLDL in these subjects (Havel and Kane, 1973). As discussed above, apoE is a glycoprotein which contains at least four different polymorphic forms designated apoE-I, -II, -III, and -IV having isoelectric points of 5.3, 5.5, 5.6 and 5.75, respectively. In type III hyperlipoproteinemia, apoE-III and apoE-IV are either deficient or missing (Utermann *et al.,* 1977; Zannis and Breslow, 1980). Havel *et al.* (1980) have recently shown that the estrogen-stimulated hepatic uptake of apoE–phospholipid complexes is specific for the isoforms apoE-III and -IV in the rat. These data provide evidence that apoE is an essential component of the lipoprotein particle responsible for the recognition of specific receptor sites in liver cells.

V. LIPID TRANSPORT AND ATHEROSCLEROSIS

A. Cellular Mechanisms of Atherogenesis

Based on gross microscopic appearance, atherosclerotic lesions have been classified into three morphologically distinct forms and include the fatty dot or streak, the fibrous plaque, and the complicated or calcified lesion. The fatty streak is the first type of lesion to appear in man, being present in almost all individuals by the third decade. It is typically a smooth lesion, orange in appearance, and contains lipid droplets located in the cytoplasm of smooth muscle cells. The major cholesteryl ester deposited in the streak is cholesteryl oleate (Small, 1977). The major cholesteryl ester in plasma lipoproteins is cholesteryl linoleate. Because of the oleate content of the fatty streak, it has been suggested that the LDL taken up by smooth muscle cells are catabolized, the cholesteryl esters hydrolyzed to give free cholesterol, and then ACAT reesterifies the free cholesterol to yield cholesteryl oleate (Fig. 3). In contrast to fatty streaks, fibrous plaques are whitish in gross appearance and contain primarily extracellular lipids, the major cholesteryl ester being cholesteryl linoleate. The complicated calcified lesion represents the final stage of development of the atherosclerotic plaque. In these lesions, there is crystallization of both free and esterified cholesterol and cell degeneration. What are the events which lead to development of the typical advanced fibrous lesion? As recently reviewed (Ross and Glomset, 1976; Ross, 1979), the major cause of progression of the lesion is proliferation of intimal smooth muscle cells, formation of extracellular connective tissue, and deposition of lipids both within the cell and in the extracellular space. The intimal smooth muscle cells begin to proliferate when there is injury to the endothelial cell lining of the arterial wall. This hypothesis for the initiation of atherosclerosis has been termed the response-to-injury hypothesis (Ross, 1979). As a result of injury, the smooth muscle cells of the artery migrate from the media into the intima. The proliferating smooth muscle cells then become exposed to the plasma components, platelets and lipoproteins. With repeated injury to the arterial wall, the smooth muscle cells continue to accumulate in the intima and develop into a fibrous lesion. The major unanswered questions regarding the sequence of events in the development of atherosclerosis relate to the mechanism by which smooth muscle cells take up plasma lipoproteins and the mechanisms by which cholesterol is transported out of the arterial cells. These aspects of atherogenesis are discussed in the following section.

B. Role of Lipoprotein Receptors

The pathways for the transport of lipid from sites of synthesis in the intestine and liver to sites of catabolism are shown in Fig. 4 and discussed in recent reviews (Nestel, 1980; Goldstein and Brown, 1977). During catabolism of the triglyceride-rich lipoproteins by lipoprotein lipase,

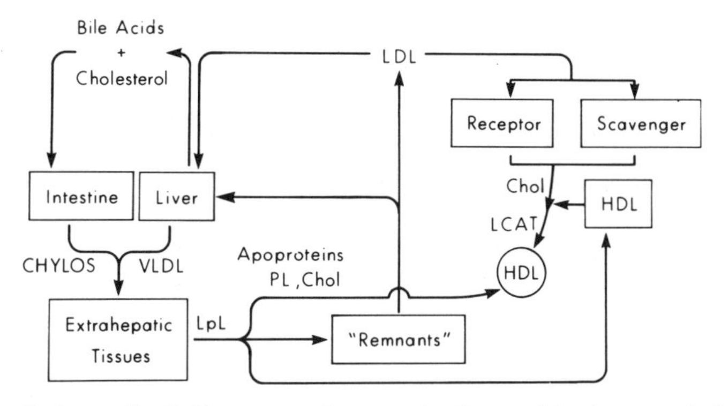

Fig. 4. Pathways for lipid transport in man. As discussed in the text, chylomicrons (CHYLOS) and VLDL are catabolized by extrahepatic tissues, yielding LDL and lipoprotein remnants which are then taken up either by the liver or by peripheral tissues. PL, phospholipid; Chol, unesterified cholesterol.

remnant lipoproteins are produced which are then either cleared from the circulation by the liver or converted to LDL. It is also possible that apoE-containing remnants can also be catabolized by peripheral tissues. Two mechanisms exist for the removal of LDL from the circulation. One involves a specific LDL receptor mechanism which recognizes either apoB- or apoE-containing lipoproteins, and the other is a scavenger pathway which is not subject to feedback regulation. The net effect of the uptake of LDL cholesteryl esters is to convert the lipid to free sterol so that it can be removed from the cell by HDL, presumably through the action of LCAT (Glomset, 1970). In addition to the uptake of lipoproteins by smooth muscle cells, several recent studies suggest that there are receptors for modified forms of LDL, particularly in macrophages (Fig. 5). As shown in Fig. 5, there are specific receptors for anionic LDL, for LDL complexed to glycosaminoglycans, for anti-LDL–LDL, and for β-VLDL isolated from plasma of cholesterol-fed dogs (Goldstein *et al.*, 1980; Basu *et al.*, 1979). With regard to anionic LDL, Fogelman *et al.* (1980) have recently shown that chemical modification of LDL with malondialdehyde increases the rate of the receptor-mediated uptake of LDL by human macrophages. Regardless of the mechanism for the uptake of LDL, the cholesteryl linoleate which accumulates in the lysosome compartment of the cell is hydrolyzed by an acid lipase, yielding free cholesterol. The cholesterol is then either removed from the cell by cholesterol receptors in the circulation or in the absence of an acceptor is converted to cholesteryl oleate and stored as a lipid within the cell. The process of cholesterol homeostasis thus involves mechanisms for the uptake of cholesterol from the circulation and for its removal from

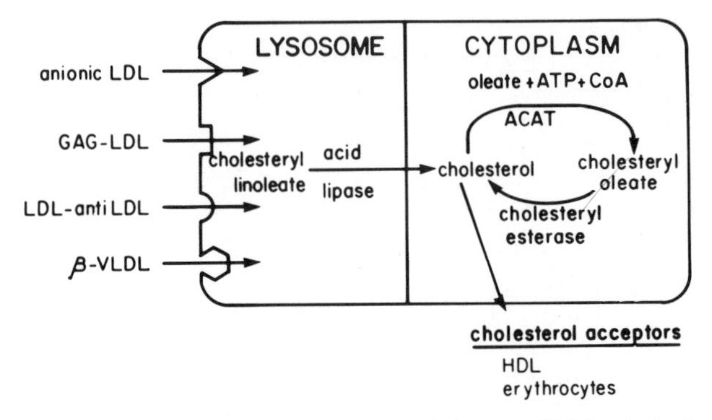

Fig. 5. Pathway for uptake of various modified forms of LDL. Anionic LDL, LDL complexed with glycosaminoglycans (GAG-LDL), anti-LDL–LDL complexes, or β-VLDL produced from cholesterol-fed animals bind to specific receptors in macrophages. The cholesteryl esters are hydrolyzed to yield free cholesterol. In the presence of a cholesterol acceptor such as HDL, the cholesterol is removed from the cell. In the absence of a cholesterol acceptor, the cholesterol is reesterified by acyl cholesterol acyltransferase to yield cholesteryl oleate.

peripheral tissues and return to the liver so that it can be secreted from the body.

C. Role of High-Density Lipoproteins

The role of HDL in regulating the cholesterol content of extrahepatic tissues has recently been reviewed by Miller (1980). It is known from epidemiological studies that low levels of plasma HDL are associated with an increased incidence of premature atherosclerosis, whereas individuals with high levels of HDL have greater life expectancies (Castelli *et al.,* 1977). As discussed above, HDL are heterogeneous and contain at least three major components (Anderson *et al.,* 1977). The HDL subfraction which correlates best with the incidence of cardiovascular disease is HDL_2. It is possible that apoproteins, particularly apoA-I and apoB, also serve as indicators for the risk of atherosclerosis (Avogaro *et al.,* 1979). The mechanism by which HDL decreases the atherosclerotic process is not entirely clear. It could function by removing free cholesterol from peripheral cells as shown in Figs. 4 and 5, or it could increase the rate of catabolism of triglyceride-rich lipoproteins by enhancing the interaction of apoC-II with lipoprotein lipase. Carew *et al.* (1976) have also suggested that HDL may inhibit the uptake and degradation of LDL in smooth muscle cells. In spite of the uncertainties concerning the mechanisms by which HDL affect the atherosclerotic process, it seems clear that individuals with higher levels of HDL are protected against atherosclerosis. Clearly, more biochemical studies are required to delineate these mechanisms.

VI. SUMMARY

In this chapter, recent concepts in lipid metabolism as they relate to the delivery of cholesterol from sites of synthesis to peripheral cells, particularly to arterial smooth muscle cells, for catabolism have been reviewed. It is now clear that plasma apolipoproteins play important roles in lipoprotein metabolism and in the delivery of cholesterol to the cell membrane. The binding of specific apoproteins, apoB and apoE, to specific cell surface receptors provides recognition sites for the uptake of plasma lipoproteins. When these regulatory mechanisms are disrupted, as is the case in the development of atherosclerosis, there is a massive accumulation of cholesteryl esters within the arterial smooth muscle cells. Clearly, if we are to understand the control mechanisms for the delivery of cholesterol to peripheral cells, we must understand the roles the various apolipoproteins play in lipoprotein metabolism. Some of these roles have been delineated, especially with respect to apoA-I, apoC-II, apoE, and apoB. Almost nothing is known concerning the roles of the other apolipoproteins. The physiology of the arterial smooth muscle cell is undoubtedly regulated by the amount and kinds of lipids delivered to the cells by the plasma lipoproteins. Alteration of the arterial smooth cell's membrane by increasing the cholesterol content or by increasing or decreasing the saturation or unsaturation of the membrane phospholipids may alter the physiology of the cell, leading to the pathogenesis of atherosclerosis.

ACKNOWLEDGMENTS

Work from the author's laboratory described in this chapter was supported in part by grants from the National Institutes of Health (HL 22619, HL 23019, and HL 24744). The author gratefully acknowledges the editorial assistance of Mrs. Janet Simons and Mrs. Beatrice Menkhaus in the preparation of the manuscript, and Ms. Gwen Kraft in preparing the artwork.

REFERENCES

Albers, J. J., Cabana, V. G., and Stahl, Y. D. B. (1976). Purification and characterization of human plasma lecithin:cholesterol acyltransferase. *Biochemistry* **15**, 1084–1087.

Anderson, D. W., Nichols, A. V., Forte, T. M., and Lindgren, F. T. (1977). Particle distribution of human serum high density lipoproteins. *Biochim. Biophys. Acta* **493,** 55–68.

Anderson, D. W., Nichols, A. V., Pan, S. S., and Lindgren, F. T. (1978). High-density lipoprotein distribution: Resolution and determination of three major components in a normal population sample. *Atherosclerosis* **29,** 161–179.

Aron, L., Jones, S., and Fielding, C. J. (1978). Human plasma lecithin:cholesterol acyltransferase: Characterization of cofactor-dependent phospholipase activity. *J. Biol. Chem.* **253,** 7220–7226.

Atkinson, D., Deckelbaum, R. J., Small, D. M., and Shipley, G. G. (1977). Structure of human plasma low-density lipoproteins: Molecular organization of the central core. *Proc. Natl. Acad. Sci. U.S.A.* **74,** 1042–1046.

Avila, E. M., Hamilton, J. A., Harmony, J. A. K., Allerhand, A., and Cordes, E. H. (1978). Natural abundance ^{13}C nuclear magnetic resonance studies of human plasma high density lipoproteins. *J. Biol. Chem.* **253,** 3983–3987.

Avogaro, P., Bon, G. B., Cazzolato, G., and Quinci, G. B. (1979). Are apolipoproteins better discriminators than lipids for atherosclerosis? *Lancet* **1,** 901–903.

Baker, H. N., Delahunty, T., Gotto, A. M., and Jackson, R. L. (1974). The primary structure of high density apolipoprotein-glutamine-I. *Proc. Natl. Acad. Sci. U.S.A.* **71,** 3631–3634.

Barter, P. J., and Lally, J. I. (1979). *In vitro* exchanges of esterified cholesterol between serum lipoprotein fractions: Studies of humans and rabbits. *Metab. Clin. Exp.* **28,** 230–236.

Basu, S. K., Brown, M. S., Ho, Y. K., and Goldstein, J. L. (1979). Degradation of low density lipoprotein dextran sulfate complexes associated with deposition of cholesteryl esters in mouse macrophages. *J. Biol. Chem.* **254,** 7141–7146.

Bengtsson, G., and Olivecrona, T. (1977). Interaction of lipoprotein lipase with heparin-Sepharose: Evaluation of conditions for affinity binding. *Biochem. J.* **167,** 109–119.

Berman, M., Hall, M., Levy, R. I., Eisenberg, S., Bilheimer, D. W., Phair, R. D., and Goebel, R. H. (1978). Metabolism of apoB and apoC lipoproteins in man: Kinetic studies in normal and hyperlipoproteinemic subjects. *J. Lipid Res.* **19,** 38–56.

Blum, C. B., Levy, R. I., Eisenberg, S., Hall, M., Goebel, R. H., and Berman, M. (1977). High density lipoprotein metabolism in man. *J. Clin. Invest.* **60,** 795–807.

Bradley, W. A., and Gotto, A. M. (1978). Structure of intact human plasma lipoproteins. *In* "Distrubances in Lipid and Lipoprotein Metabolism" (J. M. Dietschy, A. M. Gotto, and J. A. Ontko, eds.), pp. 111–137. Am. Physiol. Soc., Bethesda, Maryland.

Bradley, W. A., Rohde, M. F., Gotto, A. M., and Jackson, R. L. (1978). The cyanogen bromide peptides of the apoprotein of low-density lipoprotein (apoB): Its molecular weight from a chemical view. *Biochem. Biophys. Res. Commun.* **81,** 928–935.

Brewer, H. B., Lux, S. E., Ronan, R., and John, K. M. (1972). Amino acid sequence of human apoLp-Gln-II (apoA-II), an apolipoprotein isolated from the high-density lipoprotein complex. *Proc. Natl. Acad. Sci. U.S.A.* **69,** 1304–1308.

Brewer, H. B., Fairwell, T. LaRue, A., Ronan, R., Houser, A., and Bronzert, T. J. (1978). The amino acid sequence of human apoA-I, an apolipoprotein ioslated from high density lipoproteins. *Biochem. Biophys. Res. Commun.* **80,** 623–630.

Brown, M. S., and Goldstein, J. L. (1979). Receptor-mediated endocytosis: Insights from the lipoprotein receptor system. *Proc. Natl. Acad. Sci. U.S.A.* **76,** 3330–3337.

Carew, T. E., Hayes, S. B., Koschinsky, T., and Steinberg, D. (1976). A mechanism by which high-density lipoproteins may slow the atherogenic process. *Lancet* **1,** 1315–1317.

Castelli, W. P., Doyle, J. T., Gordon, T., Hames, C. G., Hjortland, M. C., Hulley, S. B.,

Kagen, A., and Zukel, W. J. (1977). HDL cholesterol and other lipids in coronary heart disease: The cooperative lipoprotein phenotyping study. *Circulation* **55,** 767-772.

Catapano, A. L., Kinnunen, P. K. J., Breckenridge, W. C., Gotto, A. M., Jackson, R. L., Little, J. A., Smith, L. C., and Sparrow, J. T. (1979). Lipolysis of apoC-II deficient very low density lipoproteins: Enhancement of lipoprotein lipase action by synthetic fragments of apoC-II. *Biochem. Biophys. Res. Commun.* **89,** 951-957.

Chajek, T., and Fielding, C. J. (1978). Isolation and characterization of a human serum cholesteryl ester transfer protein. *Proc. Natl. Acad. Sci. U.S.A.* **75,** 3445-3449.

Chan, L., Jackson, R. L., and Means, A. R. (1978). Regulation of lipoprotein synthesis: Studies on the molecular mechanisms of lipoprotein synthesis and their regulation by estrogen in the cockerel. *Circ. Res.* **43,** 209-217.

Chao, Y., Windler, E. E., Chen, G. C., and Havel, R. J. (1979). Hepatic catabolism of rat and human lipoproteins in rats treated with 17α-ethinyl estradiol. *J. Biol. Chem.* **254,** 11360-11366.

Chung, J., Abano, D. A., Fless, G. M., and Scanu, A. M. (1979). Isolation, properties, and mechanism of *in vitro* action of lecithin:cholesterol acyltransferase from human plasma. *J. Biol. Chem.* **254,** 7456-7464.

Cox, D. W., Breckenridge, W. C., and Little, J. A. (1978). Inheritance of apolipoprotein C-II deficiency with hypertriglyceridemia and pancreatitis. *N. Engl. J. Med.* **299,** 1421-1424.

Deckelbaum, R. J., Shipley, G. G., Small, D. M., Lees, R. S., and George, P. K. (1975). Thermal transitions in human plasma low density lipoproteins. *Science* **190,** 392-394.

Deckelbaum, R. J., Shipley, G. G., and Small, D. M. (1977a). Structure and interactions of lipids in human plasma low density lipoproteins. *J. Biol. Chem.* **252,** 744-754.

Deckelbaum, R. J., Tall, A. R., and Small, D. M. (1977b). Interaction of cholesterol ester and triglyceride in human plasma very low density lipoprotein. *J. Lipid Res.* **18,** 164-168.

Deutsch, D. G., Heinrikson, R. L., Foreman, J., and Scanu, A. M. (1978). Studies of the cyanogen bromide fragments of the apoprotein of human serum low density lipoproteins *Biochim. Biophys. Acta* **529,** 342-450.

Edelstein, C., Kezdy, F. J., Scanu, A. M., and Shen, B. W. (1979). Apolipoproteins and the structureal organization of plasma lipoproteins: Human plasma high density lipoprotein-3. *J. Lipid Res.* **20,** 143-153.

Falko, J. M., Schonfeld, G., Witzturn, J. L., Kolar, J. B., and Salmon, P. (1980). Effects of short-term high carbohydrate, fat-free diet on plasma levels of apoC-II and apoC-III and on the apoC subspecies in human plasma lipoproteins. *Metab. Clin. Exp.* **29,** 654-661.

Felker, T. E., Fainaru, M., Hamilton, R. L., and Havel, R. J. (1977). Secretion of the arginine-rich and A-I apolipoproteins by the isolated perfused rat liver. *J. Lipid Res.* **18,** 465-473.

Felts, J. M., Itakura, H., and Crane, R. T. (1975). The mechanism of assimilation of constituents of chylomicrons, very low density lipoproteins and remnants—A new theory. *Biochem. Biophys. Res. Commun.* **66,** 1467-1475.

Fielding, C. J., and Fielding, P. E. (1976). Chylomicron protein content and the rate of lipoprotein lipase activity. *J. Lipid Res.* **17,** 419-423.

Fielding, C. J., and Havel, R. J. (1977). Lipoprotein lipase. *Arch. Pathol. Lab. Med.* **101,** 225-229.

Fielding, C. J., Shore, V. G., and Fielding, P. E. (1972). A protein cofactor of lecithin:cholesterol acyltransferase. *Biochem. Biophys. Res. Commun.* **46,** 1493-1498.

Fogelman, A. M., Schechter, I., Seager, J., Hokom, M., Child, J. S., and Edwards, P. A. (1980). Malondialdehyde alteration of low density lipoproteins leads to cholesteryl ester accumulation in human monocyte-macrophages. *Proc. Natl. Acad. Sci. U.S.A.* **77,** 2214-2218.

Forte, T. M., Krauss, R. M., Lindgren, F. T., and Nichols, A. V. (1979). Changes in plasma lipoprotein distribution and formation of two unusual particles after heparin-induced lipolysis in hypertriglyceridemic subjects. *Proc. Natl. Acad. Sci. U.S.A.* **76,** 5934-5938.

Glickman, R. M., and Green, P. H. R. (1977). The intestine as a source of apolipoprotein A_1. *Proc. Natl. Acad. Sci. U.S.A.* **74,** 2569-2573.

Glickman, R. M., Green, P. H. R., Lees, R. S., and Tall, A. (1978). Apoprotein A-I synthesis in normal intestinal mucosa and in Tangier disease. *N. Engl. J. Med.* **299,** 1424-1427.

Glickman, R. M., Green, P. H. R., Lees, R. S., Lux, S. E., and Kilgore, A. (1979). Immunofluorescence studies of apolipoprotein B in intestinal mucosa: Absence in abetalipoproteinemia. *Gastroenterology* **76,** 288-292.

Glomset, J. A. (1970). Physiological role of lecithin–cholesterol acyltransferase. *Am. J. Clin. Nutr.* **23,** 1129-1136.

Goldstein, J. L., and Brown, M. S. (1977). The low density lipoprotein pathway and its relation to atherosclerosis. *Annu. Rev. Biochem.* **46,** 897-930.

Goldstein, J. L., Anderson, R. G. W., and Brown, M. S. (1979). Coated pits, coated vesicles, and receptor-mediated endocytosis. *Nature (London)* **279,** 679-685.

Goldstein, J. L., Ho, Y. K., Brown, M. S., Innerarity, T. L., and Mahley, R. W. (1980). Cholesteryl ester accumulation in macrophages resulting from receptor-mediated uptake and degradation of hypercholesterolemic canine very low density β-lipoproteins. *J. Biol. Chem.* **255,** 1839-1848.

Green, P. H. R., Tall, A. R., and Glickman, R. M. (1978). Rat intestine secretes discoid high density lipoprotein *J. Clin. Invest.* **61,** 528-534.

Grow, T. E., and Fried, M. (1978). Interchange of apoprotein components between the human plasma high density lipoprotein subclasses HDL_2 and HDL_3 *in vitro. J. Biol. Chem.* **253,** 8034-8041.

Gulik-Krzywicki, T., Yates, M., and Aggerbeck, L. P. (1979). Structure of serum low-density lipoprotein. II. A freeze-etching electron microscopy study. *Mol. Biol.* **131,** 475-484.

Hamilton, J. A., and Cordes, E. H. (1978). Molecular dynamics of lipids in human plasma high density lipoproteins: A high field ^{13}C NMR study. *J. Biol. Chem.* **253,** 5193-5198.

Hamilton, R. L., Williams, M. C., Fielding, C. J., and Havel, R. J. (1976). Discoidal bilayer structure of nascent high density lipoproteins from perfused rat liver. *J. Clin. Invest.* **58,** 667-680.

Havel, R. J., and Kane, J. P. (1973). Primary dysbetalipoproteinemia: Predominance of a specific apoprotein species in triglyceride-rich lipoproteins. *Proc. Natl. Acad. Sci. U.S.A.* **70,** 2015-2019.

Havel, R. J., Kane, J. P., and Kashyap, M. L. (1973). Interchange of apolipoproteins between chylomicrons and high density lipoproteins during alimentary lipemia in man. *J. Clin. Invest.* **52,** 32-38.

Havel, R. J., Chao, Y., Windler, E. E., Kotite, L., and Guo, L. S. S. (1980). Isoprotein specificity in the hepatic uptake of apolipoprotein E and the pathogenesis of familial dysbetalipoproteinemia. *Proc. Natl. Acad. Sci. U.S.A.* **77,** 4349-4353.

Henderson, T. O., Kruski, A. W., Davis, L. G., Glonek, T., and Scanu, A. M. (1975). ^{31}P Nuclear magnetic resonance studies on serum low and high density lipoproteins: Effect of paramagnetic ion. *Biochemistry* **14,** 1915-1920.

Ihm, J., Harmony, J. A. K., Ellsworth, J., and Jackson, R. L. (1980). Simultaneous transfer of cholesteryl ester and phospholipid by protein(s) isolated from human lipoprotein-free plasma. *Biochem. Biophys. Res. Commun.* **93,** 1114-1120.

Jackson, R. L., Morrisett, J. D., and Gotto, A. M. (1976). Lipoprotein structure and metabolism. *Physiol. Rev.* **56,** 259-316.

Jackson, R. L., Pattus, F., and Demel, R. A. (1979). Interaction of plasma apolipoproteins with lipid monolayers. *Biochim Biophys. Acta* **556,** 369-387.

Jackson, R. L., Pattus, F., and de Haas, G. (1980). Mechanism of action of milk lipoprotein lipase at substrate interfaces: Effects of apolipoproteins. *Biochemistry* **19,** 373-378.

Jonas, A. (1977). Microviscosity of lipid domains in human serum lipoproteins. *Biochim. Biophys. Acta* **486,** 10-22.

Kane, J. P., Hardman, D. A., and Paulus, H. E. (1980). Heterogeneity of apolipoprotein B: Isolation of a new species from human chylomicrons. *Proc. Natl. Acad. Sci. U.S.A.* **77,** 2465-2469.

Kashyap, M. L., Srivastava, L. S., Chen, C. Y., Perisutti, G., Campbell, M., Lutmer, R. F., and Glueck, C. J. (1977). Radioimmunoassay of human apolipoprotein C-II: A study in normal and hypertriglyceridemic subjects. *J. Clin. Invest.* **60,** 171-180.

Kashyap, M. L., Srivastava, L. S., Hynd, B. A., Perisutti, G., Brady, D. W., Gartside, P., and Glueck, C. J. (1978). The role of high density lipoprotein apolipoprotein C-II in triglyceride metabolism. *Lipids* **13,** 933-942.

Kinnunen, P. K. J., Jackson, R. L., Smith, L. C., Gotto, A. M., and Sparrow, J. T. (1977). Activation of lipoprotein lipase by native and synthetic fragments of human plasma apolipoprotein C-II. *Proc. Natl. Acad. Sci. U.S.A.* **74,** 4848-4851.

Kirchhausen, T., Fless, G., and Scanu, A. M. (1980). The structure of plasma low density lipoproteins: Experimental facts and interpretations—A minireview. *Lipids* **15,** 464-467.

Krauss, R. M., Levy, R. I., and Fredrickson, D. S. (1974). Selective measurement of two lipase activities in postheparin plasma from normal subjects and patients with hyperlipoproteinemia. *J. Clin. Invest.* **54,** 1107-1124.

Laggner, P., Muller, K., Kratky, O., Kostner, G., and Holasek, A. (1973). Studies on the structure of lipoprotein A of human high density lipoprotein HDL_3: The spherically averaged electron density distribution. *FEBS Lett.* **33,** 77-80.

Luzzati, V., Tardieu, A., and Aggerbeck, L. P. (1979). Structure of serum low-density lipoprotein. I. A solution X-ray scattering study of a hyperlipidemic monkey low-density lipoprotein. *J. Mol. Biol.* **131,** 435-473.

McConathy, W. J., and Alaupovic, P. (1976). Studies on the isolation and partial characterization of apolipoprotein D and lipoprotein D of human plasma. *Biochemistry* **15,** 515-520.

Massey, J. B., Gotto, A. M., and Pownall, H. J. (1979). Contribution of α helix formation in human plasma apolipoproteins to their enthalpy of association with phospholipids. *J. Biol. Chem.* **254,** 9359-9361.

Miller, A. L., and Smith, L. C. (1973). Activation of lipoprotein lipase by apolipoprotein glutamic acid. *J. Biol. Chem.* **248,** 3359-3362.

Miller, G. J. (1980). High density lipoproteins and atherosclerosis. *Annu. Rev. Med.* **31,** 97-108.

Morrisett, J. D., Jackson, R. L., and Gotto, A. M. (1977). Lipid-protein interactions in the plasma lipoproteins. *Biochim. Biophys. Acta* **472,** 93-133.

Nestel, P. J. (1980). Lipoprotein protein receptors and their relation to atherosclerosis. *Circ. Res.* **46** Suppl. I, 106-109.

Nichols, A. V., and Smith, L. (1965). Effect of very low density lipoproteins on lipid transfer in incubated serum. *J. Lipid Res.* **6,** 206-210.

Osborne, J. C., and Brewer, H. B. (1977). The plasma lipoproteins. *Adv. Protein Chem.* **31,** 253–337.

Patsch, J. R., Gotto, A. M., Olivecrona, T., and Eisenberg, S. (1978). *Proc. Natl. Acad. Sci. U.S.A.* **75,** 4519–4523.

Pattnaik, N. M., Kezdy, F. J., and Scanu, A. M. (1976). Kinetic study of the action of snake venom phospholipase A_2 on human serum high density lipoprotein 3. *J. Biol. Chem.* **251,** 1984–1990.

Pattnaik, N. M., Montes, A., Hughes, L. B., and Zilversmit, D. B. (1978). Cholesteryl ester exchange protein in human plasma isolation and characterization. *Biochim. Biophys. Acta* **530,** 428–438.

Pitas, R. E., Innerarity, T. L., Arnold, K. S., and Mahley, R. W. (1979). Rate and equilibrium constants for binding of apo-E HDL_c (a cholesterol-induced lipoprotein) and low density lipoproteins to human fibroblasts: Evidence for multiple receptor binding of apo-E HDL_c. *Proc. Natl. Acad. Sci. U.S.A.* **76,** 2311–2315.

Pownall, H. J., Hu, A., Gotto, A. M., Albers, J. J., and Sparrow, J. T. (1980). Activation of lecithin:cholesterol acyltransferase by a synthetic model lipid-associating peptide. *Proc. Natl. Acad. Sci. U.S.A.* **77,** 3154–3158.

Quarfordt, S., Hanks, J., Jones, R. S., and Shelburne, F. (1980). The uptake of high density lipoprotein cholesteryl ester in the perfused rat liver. *J. Biol. Chem.* **255,** 2934–2937.

Ross, R. (1979). The arterial wall and atherosclerosis. *Annu. Rev. Med.* **30,** 1–15.

Ross, R., and Glomset, J. A. (1976). The pathogenesis of atherosclerosis. *N. Engl. J. Med.* **295,** 369–377, 420–425.

Schaefer, E. J., Blum, C. B., Levy, R. I., Jenkins, L. L., Alaupovic, P., Foster, D. M., and Brewer, H. B. (1978a). Metabolism of high-density lipoprotein apolipoproteins in Tangier disease. *N. Engl. J. Med.* **299,** 905–910.

Schaefer, E. J., Eisenberg, S., and Levy, R. I. (1978b). Lipoprotein apoprotein metabolism. *J. Lipid Res.* **19,** 667–687.

Schaefer, E. J., Jenkins, L. L., and Brewer, H. B. (1978c). Human chylomicron apolipoprotein metabolism. *Biochem. Biophys. Res. Commun.* **80,** 405–412.

Schaefer, E. J., Zech, L. A., Schwartz, D. E., and Brewer, H. B. (1980). Coronary heart disease prevalence and other clinical features in familial high-density lipoprotein deficiency (Tangier disease). *Ann. Intern. Med.* **93,** 261–266.

Schneider, H., Morrod, R. S., Colvin, J. R., and Tattrie, N. H. (1973). The lipid core model of lipoproteins. *Chem. Phys. Lipids* **10,** 328–353.

Schonfeld, G., Bell, E., and Alpers, D. H. (1978). Intestinal apoproteins during fat absorption. *J. Clin. Invest.* **61,** 1539–1550.

Schonfeld, G., Grimme, N., and Alpers, D. (1980). Detection of apolipoprotein C in human and rat enterocytes. *J. Cell Biol.* **86,** 562–567.

Scow, R. O., Desnuelle, P., and Verger, R. (1979). Lipolysis and lipid movement in a membrane model: Action of lipoprotein lipase. *J. Biol. Chem.* **254,** 6456–6463.

Scow, R. O., Blanchette-Mackie, E. J., and Smith, L. C. (1980). Transport of lipid across capillary endothelium. *Fed. Proc., Fed. Am. Soc. Exp. Biol.* **39,** 2610–2617.

Segrest, J. P. (1977). Amphipathic helixes and plasma lipoproteins: Thermodynamic and geometric considerations. *Chem. Phys. Lipids* **18,** 7–22.

Segrest, J. P., Jackson, R. L., Morrisett, J. D., and Gotto, A. M. (1974). A molecular theory of lipid-protein interactions in the plasma lipoproteins. *FEBS Lett.* **38,** 247–253.

Shelburne, F. A., and Quarfordt, S. H. (1974). A new apoprotein of human very low density lipoproteins. *J. Biol. Chem.* **249,** 1428–1433.

Shelburne, F. A., and Quarfordt, S. H. (1977). The interaction of heparin with an apoprotein of human very low density lipoprotein. *J. Clin. Invest.* **60,** 944–950.

Shen, B. W., and Scanu, A. M. (1980). Properties of human apolipoprotein A-I at the air-water interface. *Biochemistry* **19,** 3643-3650.

Sherrill, B. C., Innerarity, T. L., and Mahley, R. W. (1980). Rapid hepatic clearance of the canine lipoproteins containing only the E apoprotein by a high-affinity receptor. *J. Biol. Chem.* **255,** 1804-1807.

Shipley, G. G., Atkinson, D., and Scanu, A. M. (1972). Small-angle X-ray scattering of human serum high-density lipoproteins. *J. Supramol. Struct.* **1,** 98-104.

Shore, V. G., and Shore, B. (1973). Heterogeneity of human plasma very low density lipoproteins: Separation of species differing in protein components. *Biochemistry* **12,** 502-507.

Small, D. M. (1977). Cellular mechanisms for lipid deposition in atherosclerosis. *N. Eng. J. Med.* **297,** 873-877, 924-929.

Smith, L. C., Pownall, H. J., and Gotto, A. M. (1978). The plasma lipoproteins: Structure and metabolism. *Annu. Rev. Biochem.* **47,** 751-777.

Socorro, L., and Camejo, G. (1979). Preparation and properties of soluble, immunoreactive apoLDL. *J. Lipid Res.* **20,** 631-638.

Stamler, J. (1980). Data base on the major cardiovascular diseases in the United States. *Atheroscler. Rev.* **7,** 49-96.

Steele, J. C. H., and Reynolds, J. A. (1979). Molecular weight and hydrodynamic properties of apolipoprotein B in guanidine hydrochloride and sodium dodecyl sulfate solutions. *J. Biol. Chem.* **254,** 1639-1643.

Steinberg, D. (1979). Origin, Turnover, and fate of plasma low-density lipoprotein. *Prog. Biomed. Pharmacol.* **15,** 166-199.

Stern, M. P. (1979). The recent decline in ischemic heart disease mortality. *Ann. Intern. Med.* **91,** 630-640.

Tall, A. R., and Small, D. M. (1978). Plasma high-density lipoproteins. *N. Engl. J. Med.* **279,** 1232-1236.

Tall, A. R., Shipley, G. G., and Small, D. M. (1976). Conformational and thermodynamic properties of apoA-I of human plasma high-density lipoproteins. *J. Biol. Chem.* **251,** 3749-3755.

Tall, A. R., Small, D. M., Deckelbaum, R. J., and Shipley, G. G. (1977a). Structure and thermodynamic properties of high density lipoprotein recombinants. *J. Biol. Chem.* **252,** 4701-4711.

Tall, A. R., Atkinson, D., Small, D. M., and Mahley, R. W. (1977b). Characterization of the lipoproteins of atherosclerotic swine. *J. Biol. Chem.* **252,** 7288-7293.

Tall, A. R., Deckelbaum, R. J., Small, D. M., and Shipley, G. G. (1977c). Thermal behavior of human plasma high density lipoprotein. *Biochim. Biophys. Acta* **487,** 145-153.

Tall, A. R., Small, D. M., Atkinson, D., and Rudel, L. L. (1978). Studies on the structure of low density lipoproteins isolated from *Macaca fasicularis* fed an atherogenic diet. *J. Clin. Invest.* **62,** 1354-1363.

Utermann, G., Hees, M., and Steinmetz, A. (1977). Polymorphism of apolipoprotein E and occurrence of dysbetalipoproteinaemia in man. *Nature (London)* **269,** 604-607.

Van Berkel, T. J. C., Kruijt, J. K., Van Gent, T., and Van Tol, A. (1980). Saturable high affinity binding of low density and high density lipoprotein by parenchymal and non-parenchymal cells from rat liver. *Biochem. Biophys. Res. Commun.* **92,** 1002-1008.

Weisgraber, K. H., and Mahley, R. W. (1978). Apoprotein (E-A-II) complex of human plasma lipoproteins. I. Characterization of this mixed disulfide and its identification in a high density lipoprotein subfraction. *J. Biol. Chem.* **253,** 6281-6288.

Weisgraber, K. H., Troxler, R. F., Rall, S. C., and Mahley, R. W. (1980). Comparison of the human, canine, and swine E apoproteins. *Biochem. Biophys. Res. Commun.* **95,** 374-380.

Windler, E., Chao, Y., and Havel, R. J. (1980a). Determinants of hepatic uptake of triglyceride-rich lipoproteins and their remnants in the rat. *J. Biol. Chem.* **255,** 5475–5480.

Windler, E., Chao, Y., and Havel, R. J. (1980b). Regulation of the hepatic uptake of triglyceride-rich lipoproteins in the rat: Opposing effects of homologous apolipoprotein E and individual C apoproteins. *J. Biol. Chem.* **255,** 8303–8307.

Wlodawer, A., Segrest, J. P., Chung, B. H., Chiovetti, R., and Weinstein, J. N. (1979). High-density lipoprotein recombinants: Evidence for a bicycle tire micelle structure obtained by neutron scattering and electron microscopy. *FEBS Lett.* **104,** 231–235.

Yeagle, P. L., Langdon, R. G., and Martin, R. B. (1977). Phospholipid-protein interactions in human low density lipoprotein detected by ^{31}P nuclear magnetic resonance. *Biochemistry* **16,** 3487–3491.

Yokoyama, S., Fukushima, D., Kupferberg, J. P., Kezdy, F. J., and Kaiser, E. T. (1980). The mechanism of activation of lecithin:cholesterol acyltransferase by apolipoprotein A-I and an amphiphilic peptide. *J. Biol. Chem.* **255,** 7333–7339.

Zannis, V. I., and Breslow, J. L. (1980). Characterization of a unique human apolipoprotein E variant associated with type III hyperlipoproteinemia. *J. Biol. Chem.* **255,** 1759–1762.

Zilversmit, D. B., Hughes, L. B., and Balmer, J. (1975). Stimulation of cholesterol ester exchange by lipoprotein-free rabbit plasma. *Biochim. Biophys. Acta* **409,** 393–398.

Subject Index

B

C

H

I

J

K

L

N

O

T

V

W